AMBITIOUS SCIENCE TEACHING

Mark Windschitl, Jessica Thompson,
and Melissa Braaten

HARVARD EDUCATION PRESS
Cambridge, Massachusetts

Third Printing, 2018

Copyright © 2018 by the President and Fellows of Harvard College

All rights reserved. No part of this publication may be reproduced or transmitted in any form or by any means, electronic or mechanical, including photocopy, recording, or any information storage and retrieval systems, without permission in writing from the publisher.

Paperback ISBN 978-1-68253-162-4
Library Edition ISBN 978-1-68253-163-1

Library of Congress Cataloging-in-Publication Data
Names: Windschitl, Mark, author. | Thompson, Jessica Jane, author. | Braaten, Melissa L., author.
Title: Ambitious science teaching / Mark Windschitl, Jessica Thompson, and Melissa Braaten.
Description: Cambridge, Massachusetts : Harvard Education Press, 2018. | Includes bibliographical references and index.
Identifiers: LCCN 2017055920| ISBN 9781682531624 (pbk.) | ISBN 9781682531631 (library edition)
Subjects: LCSH: Science—Study and teaching—United States. | Action research in education. | Teaching—Methodology.
Classification: LCC QC183.3 .W56 2018 | DDC 507.1/2—dc23
LC record available at https://lccn.loc.gov/2017055920

Published by Harvard Education Press,
an imprint of the Harvard Education Publishing Group

Harvard Education Press
8 Story Street
Cambridge, MA 02138

Cover Design: Wilcox Design
Cover Image: kali9/E+/Getty Images
Figures 1.1, 2.1, 5.1, 8.1, and 12.1: iStock.com/skeeg

The typefaces used in this book are Latienne, Agenda, and Avenir Next.

Contents

Preface

Ambitious Science Teaching (AST) represents a vision for changing how children learn about the natural world. It focuses on the ideas and other diverse resources that they bring to classrooms every day, as building blocks for sense making and progressive knowledge building. This vision is built on a repertoire of teaching practices that cultivate student dialogue, community reasoning, and intellectual rigor, as well as the ability to learn how to learn. As promising as this sounds, we would prefer that when you read each chapter of this book, you be critical of our theories and tools—to be sure that the evidence is clear to you and is compelling enough for you to consider changes in your teaching. To accommodate this, we now share the origin story of AST, including the fits and starts that made the journey interesting.

Many years ago, long before the idea of Ambitious Science Teaching started to take shape, we were reading everything we could get our hands on about instruction that had significant impacts on student learning. Researchers in diverse fields of study were starting to describe how, under the right circumstances, children in science classrooms could explain, model, argue, design investigations, and problem-solve with one another, in ways that went far beyond the expectations built into common curriculum and standards. We became excited about the possibility of translating and applying outcomes from these studies, largely done under controlled conditions, to the dynamic environments in which science teachers work.

We were not the only ones trying to make these connections, but we felt confident about our prospects, in part because we had spent years as science educators ourselves and understood the challenges of teaching in underresourced schools with precious little time to experiment with peers about instruction. All three of us were at the University of Washington and in charge of preparing novice teachers for work in secondary science classrooms. We believed that our own courses on methods of instruction could be incubators for testing out innovative forms of practice and then seeing how these worked with young learners in local schools. Our preservice teachers were eager to learn about alternatives to the status quo, so we obliged them and spent months explaining in great detail what research indicated they should be doing to foster student engagement and authentic science activity. We later followed these novices into their host classrooms to observe the fruits of our labor. We were all disappointed.

Our exuberance about adventurous teaching and our novices' willingness to try out new and unfamiliar routines did not translate into eager student participation or the learning outcomes that the research literature had promised. Most of the exasperated teachers-in-training said things like "I knew *what* I wanted to have happen, but didn't know *how* to make it happen."

We realized then that we had relied on broad notions like "inquiry" and "hands-on work" to shape their attempts at teaching, and had failed to show our novices actual practices—that is, approaches that you could see and hear someone using in a classroom on a regular basis. What was needed were professional routines that were recognizable, principled, and improvable. This realization prompted us to specify instructional practices from the research literature, which required some inventiveness because they were not clearly identified there. We ended up selecting a small number that appeared critical for student learning and participation, gave them names, and aimed to get our novices proficient at them. Each of these practices was really a combination of tasks, talk, and tools that had to be used together to support knowledge building. At this point, we didn't want to repeat our earlier mistakes by simply *telling* our novice teachers what powerful practice was, we had to *immerse* them in it. So, we played the role of the teacher while our novices became the students. They then took their turn in the teaching role during what we and others call *rehearsals* (that's another book). In retrospect, we can clearly see that there is no substitute for live action in which our own instruction becomes public, responsive to the science ideas of everyone in the room, and open to critique. Weak spots in teaching, ours and theirs, became glaringly obvious, but the benefits were that everyone learned quickly what was possible, and how to improve. To our delight, these practices "traveled" much more readily into K–12 classrooms.

We are oversimplifying the story, but these events started a twelve-year run of experimenting with our novices, and increasingly with local teachers, around a set of core practices and the tools to go with them. These resources now get "tested" by hundreds of colleagues—both experienced and preservice educators—on a daily basis.

Other surprises followed. The core practices were originally designed for middle and high school instruction, but elementary teachers soon began pushing the limits of AST with five- and six-year-olds. We were astonished to see what was happening in their classrooms. Teachers had to adapt most of our tools, using fewer words and more pictures, but even the youngest of learners, they found, were capable of experimenting, making sense of data, and revising explanations over time. We recently observed boys and girls in a second-grade classroom create different claims for why a nearby town was nearly wiped out by a flash flood, despite only modest rainfall in the area. The diversity of their initial ideas was impressive enough, but then a few days later these young learners evaluated their claims using maps, evidence from readings, their own "sandbox" tests, data collected by scientists, and known science facts. Although such episodes never unfold without unexpected problems and require lots of support, teaching like this is still extraordinary—and slowly but surely making appearances in classrooms around the country.

The AST community is now more focused than ever on finding ways for *all students to participate* in challenging science, to make science *compelling to wider groups of learners*, and to *provide the means for students to show what they know*. If we are going to turn a corner on how science is taught, we have to be serious about including every learner in the classroom and recognizing that children *have many legitimate ways of making sense* of the natural world. Traditional teaching does not often accommodate diverse pathways to deep understanding. This stance about equity, we believe, sustains our colleagues through a lot of hard work—hard because it requires new knowledge and skills, but also because we are now trying to teach in ways that we never experienced ourselves as learners.

We dedicate this book, then, to professional educators around the country who have taken risks to make a difference for their students. Some of them use AST, some do not. We also wish to thank our current and former research assistants and postdoctoral researchers who have shared the frustrations and the joys of this ongoing project. You have been great colleagues: David Stroupe, Carolyn Colley, Sara Hagenah, Michelle Salgado, Karin Lohwasser, Christine Chew, Hosun Kang, Christie Barchenger, Kat Laxton, Biz Wright, Soo-Yean Shim, and Jen Richards.

Mark Windschitl
Jessica Thompson
Melissa Braaten

ONE

A Vision of Ambitious Science Teaching

EVERY YEAR, IN AN INCREASING NUMBER of schools across the country, science educators come together with colleagues to plan for significant changes in the ways their students will learn science. These professionals may be motivated by new approaches to teaching that challenge their current conceptions of what's possible in classrooms, or they may be unsettled by their students' chronic disengagement. Likely a bit of both. Regardless of the call to action, when teachers make the decision to improve instruction together, they have to put two questions on the table: What do we want to get better at? How do we go about changing our practice in systematic ways?

While there are never straightforward answers to these questions, successful efforts at change are always influenced by a *coherent vision of instruction* that incorporates both rigor and equity into the work that teachers aspire to do with students. But where does such a vision come from, and how does it get translated into action that we can see and hear in classrooms? This book provides a roadmap for innovation by describing how research can inform decisions teachers make every day, to shape conditions of learning for their students.

For the past thirty years, educational researchers have been studying the circumstances under which students engage with science ideas and interact with each other as they do authentic disciplinary activity together. Countless teachers have been observed and data collected in their classrooms, many of these in high-needs schools. Important questions have been asked, such as: Who is participating? Who is not? What, exactly, are students learning? What changes in teaching or in classroom conditions result in learning benefits for students? Are young learners capable of more than what current forms of teaching can support?

In some cases, researchers identified effective teachers and documented how students responded to their classroom norms, instructional routines, and high expectations. In other cases, researchers teamed up with educators to redesign the classroom environment or construct new curricula in order to enhance students' engagement and learning. As a result of all these efforts, stories emerged from classrooms where students were consistently pushing beyond the boundaries of what we consider typical school science. Seventh graders, for example, were observed piecing together increasingly complex explanations for how genetic and environmental factors interacted to drive up the rates of diabetes in their community. Juniors in a chemistry class were developing models for the spread of ocean acidification, using historical measurements of atmospheric gases, data from solubility experiments they had run, and information from scientific journals about global warming trends. And in a kindergarten classroom, five-year-olds took turns arguing whether gravity should be added to their stories about "how someone little can bump someone big off the end of a playground slide."

What made these powerful examples of learning were not the clichéd episodes of students working excitedly with beakers and microscopes or the rhetorical drama of whole-class debates. Rather, the really remarkable things were less visible. For example, a *majority* of learners in each classroom participated in the activity and sense-making conversations each day. Students were used to *sharing their thinking* with classmates, and were expected to *revise their science ideas* over the duration of a unit. Teachers provided *scaffolding* for everything from explanation writing to critiquing someone else's idea, and many educators were notably skilled in *managing productive and equitable talk* by their students.

All these trends were promising, but like pieces of a large and complicated puzzle, they needed to be assembled to form a bigger picture. Up to this point they had not been organized in a way that helped teachers apply the research findings to plans for better instruction in their classrooms. Instead, recommendations from the literature have seemed like a patchwork of loosely connected claims that didn't address the full scope of how teachers need to work with their students in order to achieve important goals. In many cases, the advice has been too theoretical to be usable or has been translated into slogans.

This picture has now changed. With this book, we introduce *Ambitious Science Teaching* (AST) as a coherent and accessible vision of what highly effective science instruction can look like. AST represents a synthesis of recommendations from many types of research studies. The goal of AST is to help students of *all* backgrounds to deeply understand fundamental science ideas, participate in the practices of science, solve authentic problems together, and learn how to continue learning on their own. The kind of teaching required to achieve these goals is adaptive to students' needs and thinking, and maintains rigorous standards for participation and performance by everyone in the classroom. More importantly, the practical ideas that make up this vision have been tested in a wide range of classrooms, and they continue to evolve as teachers use them.

Ambitious Science Teaching, however, is not a label for just any kind of roll-up-your-sleeves "good" instruction, nor is it a list of vague prescriptions. The core of this framework is made up of four sets of teaching practices (see figure 1.1). By "practices" we mean regularly recurring teaching activities that are devoted to planning for, enacting, or reflecting on instruction. The first is a set of planning practices that are used to design challenging and coherent units of instruction; the other three sets of practices come into play as you teach the unit, and require different kinds of intellectual work by students. Each set of practices comes with its own subset of tools, tasks, and opportunities for sense-making talk.

You are likely wondering why there are only four sets of practices. The answer is this: To make fundamental shifts in teaching, you have to invest yourself in practices that make the greatest difference for your students' participation and learning. If there were dozens of random strategies involved, you would be diffusing precious time and energy becoming acquainted with all of them, and you still wouldn't be sure if or how they could be used together in your classroom. In contrast, if you focus on a core group of practices—a set that works together—you can seek mastery of a few types of interactions with students that have the greatest promise for their academic success.

The other reason for concentrating efforts on four sets of practices is that it allows groups of educators who take up AST to start speaking the same language about instruction. When educators adopt a common language about teaching, they can more readily share their experiences with these practices, describe

FIGURE 1.1 Ambitious Science Teaching framework

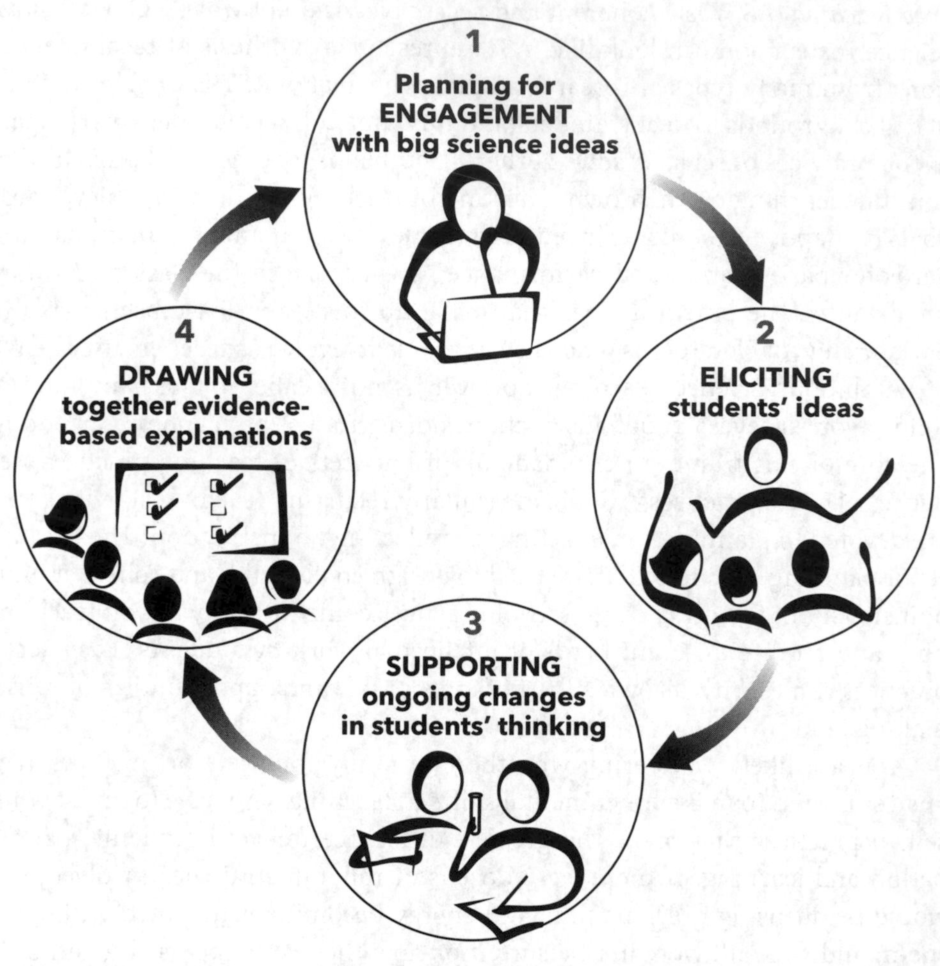

variations that work for their students, or talk about tools they've developed to support these practices. In other words, they can start to learn from each other, even as they "change" what AST looks like in their local context. So, when we refer to the continual improvement of teaching, we mean not just that individual educators become more advanced in their practice, but also that science teaching among groups of professionals is transformed over time.

The claims we make come from the fact that we've been working with hundreds of science teachers over the past twelve years on these practices. We were science educators ourselves (K–8 and high school) and are currently faculty in research universities. But the story of Ambitious Science Teaching has been written largely by our classroom colleagues. They have tested these core practices in high-needs schools with thousands of students; they have rejected some of our early tools as well intentioned but unwieldy in real instructional situations, and then created more effective ones in their place; and they have reconstructed some of our initial practices to be more effective for a wider range of students, including English language learners (ELLs) and other children on the margins of the science classroom. They continue to adapt AST while holding close to powerful research-based principles of rigorous and equitable teaching.

Our colleague-partners in the field have also *deprivatized their teaching* so that they can observe each other taking risks while working with students in unfamiliar but promising ways. This goes against the grain of traditional school culture, which has been characterized as an "egg crate" model of professional work-life—each teacher nested in his or her own classroom with little access to others who might provide glimpses of alternative and useful practices. If we aspire to the continual improvement of teaching, then opening up our classrooms to others is crucial. So the lessons we've learned about improvement efforts can be summed up this way: transformative teaching is the outcome of (1) focused work, (2) with colleagues, (3) on things that matter for student learning, (4) over time. Let's take a closer look now at what AST looks like and sounds like in classrooms.

WHAT WOULD WE SEE AND HEAR IN THE AST CLASSROOM?

We've claimed that AST is informed by research, but such assertions get tossed around a lot in the world of education, so we want to clarify what this means in the context of our framework. Because of the social dynamics of classrooms and the complexity of learning itself, no research study can ever definitively say, "Here is the key to teaching."[1] Rather, researchers analyze dozens or hundreds of similar investigations and look for trends that convince them if certain conditions are in place in classrooms, or if teachers routinely make particular kinds of moves with students, then over time there is a greater chance that more students will participate in more meaningful ways, and have more fruitful opportunities to learn. The best we can do is increase these probabilities. Sometimes our influence is in the short term (a kind of question we ask a student) and sometimes it unfolds over months (gradually building students' identities as knowers of the natural world).

So where should we place our bets? What does the research say that can be translated to teaching? We've identified seven elements of rigorous and equitable instruction by cross-referencing different bodies of research (for example, studies of how students learn, studies of expert teaching, studies of how knowledge is constructed within the discipline of science, and studies of how equity can be fostered in classrooms), and we express these as "what you'd see and hear" in an AST environment, regardless of whether it is elementary, middle, or high school. What we share next are not the practices themselves that we mentioned previously, but rather broad conditions of instruction that characterize AST; in other words, these elements form the foundations of the practices. As you read them, you'll likely notice that these elements have to work together to be effective; they cannot be used in isolation.

1. Teachers Anchor Students' Learning Experiences in Complex and Puzzling Science Phenomena[2]

For much of an instructional unit in the AST classroom, students focus on an unusual event or process in the world that they will build explanations for by coordinating different science ideas with logic and evidence. A group of fifth graders studying sound, for example, might try to explain how a singer can break a glass with just the energy from his voice. A group of ninth graders studying ecosystems might be figuring out what factors are contributing to the decline of killer whale populations and what might be done to restore balance among biotic and abiotic parts of their habitat. A class of twelfth-grade physics students might learn how forces and motion work by exploring how an urban gymnast can run up to a wall, plant his foot on it, do a backflip, and land on his feet. Students' explanations for all of these phenomena are intended to be detailed, elaborate, and backed up with multiple forms of evidence. Most, but not all, of the unit's activities are directed toward these aims.

2. Students' Hypotheses, Experiences, Cultural Knowledge, and Questions Are Treated as Resources to Help the Class Build Toward Big Science Ideas[3]

In the AST classroom, teachers work with and on students' ideas. *Working with* ideas means that teachers elicit students' initial understandings of a puzzling phenomenon or a driving question that anchors the unit. At first, students typically offer partial and fragmented explanations; they may also reference outside-of-school experiences or cultural knowledge they feel is relevant. We refer to all of these as *resources*. Occasionally these resources provide novel viewpoints that were not anticipated by the teacher, but nonetheless they provoke new kinds of reasoning by the class. The teacher's role is to prompt students to think out loud, to help them organize rough-draft ideas and represent them back to the class as "what we think we know for now," and to record questions that students are interested in. *Working on* ideas means that, throughout the unit, the teacher will invite students to use their prior academic knowledge, outside-of-school experiences, and even new questions to help the class revise and elaborate on their current explanations. Teachers keep the canonical science in mind during lessons, while at the same time listening for student contributions that have the potential to support everyone's reasoning in unexpected ways.

3. Students Use Ensembles of Scientific Practices for Testing Ideas They Believe Are Important to Their Developing Explanations and Models[4]

Students in an AST classroom pursue the long-term goal of constructing coherent and evidence-based explanations of complex phenomena. Science practices such as asking testable questions, modeling, designing and carrying out investigations, and arguing from evidence are essential to achieving this goal. Even the youngest of students can engage in these science practices if they understand the role that each plays in the larger enterprise of improving ideas, and they are apprenticed into using these practices together in order to make sense of a puzzling event. Apprenticing means that the teacher helps students participate in authentic disciplinary practices, but takes initial responsibility for the more challenging and unfamiliar aspects of the work. Over time, learners take on more central roles in the activity and decision making, with the help of feedback and other kinds of support from the teacher.

4. Teachers Provide Varied Opportunities for Students to Reason Through Talk[5]

Research studies from many different fields confirm that student talk is the primary vehicle for their sense making. Teachers engaged with rigorous and equitable instruction help their students understand the "rules of the game" for different kinds of conversations and provide tools to help them learn to participate in more productive ways over time. These tools include student-generated norms for talk (e.g., how to disagree with a peer's idea or ask for clarification), routines that structure how partners can share ideas as well as listen actively to one another, and debriefings with students about how they could improve the quality of small-group and whole-class conversations.

5. Students Have Access to Specialized Tools and Routines That Support Their Science Writing, Talk, and Participation in Activity[6]

Students in these classrooms are stretched to communicate in ways just beyond what they are currently capable of, and this is where scaffolding becomes critical for the improvement of valued performances. Scaffolding takes many forms. A teacher might, for example, provide sentence frames to help students question each other about progress during group work, or the teacher might think out loud about the steps a person should consider in designing an experiment. In some classrooms students are provided with different examples of how scientific explanations are revised based on new evidence. Young learners are routinely challenged to engage in complex performances, but these become manageable because the teacher provides just enough structure, in the right places, to do the work.

6. Student Thinking Is Made Visible and Subject to Commentary by the Classroom Community[7]

The teacher engaged in rigorous and equitable instruction periodically assembles students to do work together as a community that could not be accomplished by individuals alone. Students are asked to share knowledge products that depict their current thinking; these can be models, conceptual drawings, or partial explanations. These representations are often selected by the teacher for their potential to prompt whole-class dialogue. The conversations can then focus on one of several goals, such as comparing different ways of constructing explanations for the same event, unpacking the meaning of a science concept, or commenting on how a student's model was changed in response to new evidence. In the process, everyone has a chance to hear how peers are reasoning about problems and to pick up examples of "what counts" as a science claim, a credible argument, a causal model, or an evidence-based explanation.

7. Learning Experiences Are Selected to Help Students Build Toward Cumulative and Nuanced Understandings of Big Science Ideas[8]

Students in AST classrooms learn that their thinking is supposed to change over time—to be revised and made more coherent in response to new experiences. This is a major shift from the common schooling goal of being correct about "the idea of the day" or reproducing, by rote, knowledge that is in textbooks. In AST classrooms, students spend significant time conducting investigations, discussing readings, taking notes during interactive mini-lectures by teachers, and working on lab activities. Each of these episodes is intentionally designed to illuminate some part of a complex problem or event. Students come to expect that they'll have to integrate ideas from multiple lessons in order to genuinely understand the science.

How These Elements Work Together to Influence Opportunities to Learn

As you read about these elements, you probably noticed that they would work most effectively in combination with one another, rather than as stand-alone conditions. For example, if you want to use students' ideas and experiences as resources for learning in your classroom (second in our list), you will have to encourage open discourse (fourth in our list). If you want to make students' thinking visible so they can collectively work on one another's ideas (sixth), you'll need to structure how they present their works in progress or demonstrate how to offer public critique to a peer (fifth). And engaging in authentic disciplinary work (third) to explore ideas that change over time (seventh) is best accomplished when students are trying to figure out complex problems (first). Each element, then, is part of a larger organic experience that fosters a sense of intellectual community and challenge for students.

All of these elements also appear in various standards for professional practice and evaluation—for example, the Danielson Framework and National Board Certification rubrics (see appendix A for an analysis). Much of the same research literature that informed these documents has also been used to shape AST.

THE CENTRAL ROLE OF EQUITY IN AMBITIOUS SCIENCE TEACHING

Equity is often thought of as providing every student with similar opportunities to make academic progress. But this definition is misleading; it suggests that exposing everyone in the classroom to the same well-crafted activities or clear presentations of information qualifies as both effective and fair. What's missing? In equitable classrooms, the teacher also *provides the means by which all groups of students can take advantage of situations that are designed to support learning*.

Here's an example. We may take for granted that a captivating lab with homemade batteries will illuminate ideas about energy storage and energy transfer for every student. However, the prereading you give out about positive and negative charges may be impenetrable for learners who speak a primary language other than English, and the expectations for interacting verbally with partners about these ideas are similarly daunting—that is, unless these learners are allowed to team up with an ELL peer whose reading and speaking skills are a bit more advanced than theirs. This simple move may be all that's necessary to transform an exercise in frustration into a productive episode of learning.

Without attention to equity in instruction, some students come to believe that science is like a foreign language, convincing themselves that "I am not the kind of person who does science," or "science is just not relevant to my life." Boys and girls who struggle in the classroom do not always have learning issues; they may simply need access to basic tools, language, supports, or extra time to make challenging work more tractable. A disproportionate number of these students come from poverty-impacted, immigrant, or special needs backgrounds. For these groups in particular and all students in general, rigor is not possible without the supportive means to achieve valued goals. Because concerns for equity are so important to instructional excellence, they are woven into Ambitious Science Teaching. But we need to intentionally call out the different ways that AST supports equity so that these threads don't get lost when teachers develop variations of the core practices or create new tools for their classrooms. So let's be clear.

First, we believe that teachers should, to the degree possible, *situate learning in familiar or everyday contexts*, choosing events for study that are related to students' experiences and interests. These can be real-life science phenomena that happen in the community (e.g., local ecosystem disturbances), in homes (e.g., chemistry of food preparation), within families (e.g., how similar but non-identical physical traits appear in siblings), or at school (e.g., soundproofing the gym). These highly contextualized situations are preferred over generic textbook tasks because they allow students from diverse backgrounds to use their out-of-school knowledge as they participate in building explanations or solving problems. Students often feel additional relevance when issues of social justice are involved, such as the impact of poverty on diet or climate change's effect on vulnerable human populations.

As instruction unfolds, teachers are *responsive to students' ideas, experiences, and questions.* The teacher welcomes local knowledge that students from varied backgrounds bring to class discussions. Students' ideas are treated as legitimate mini-theories, and even their partial understandings are framed by the teacher as possible resources for the entire class to work with. Occasionally, teachers can adapt lessons to take advantage of student experiences or puzzles that they believe are relevant to the science being studied.

Because authentic science activity and talk are central to AST, *teachers make explicit to students how scientists generate and defend claims for knowing, and the norms for participation in disciplinary conversations.* Many students from non-dominant communities do not participate in science classrooms because they are unfamiliar with these "rules of the game"—meaning those valued in traditional Western science. When teachers describe ways to contribute to various kinds of discussions and model their own thinking about how to enter a conversation, they allow more students access to small-group and whole-class talk. These teacher moves reinforce norms for respectful and accountable whole-class dialogue, making students feel safe about participating. With this help, students more readily see the similarities and differences between everyday conversation and more academic conversations, and over time, recognize when to shift from one to the other.

Teachers routinely *use specialized forms of scaffolding* for reading and writing, providing structures for learners to interact with others around intellectual work, and adapting tasks for diverse learners (as we described in the previous section). These opportunities are sometimes made available to an entire class; at other times they are tailored or differentiated to meet the needs of particular students.

As instruction unfolds, *teachers honor students' sense-making repertoires.* Students from various ethnic, social class, and racial backgrounds often employ metaphors, storytelling, and physical action to express and debate sophisticated science stories. These valuable ways of reasoning and communicating are different from the predominant modes of concrete descriptions, classification talk, and stepwise explanations that are more traditionally used to express science knowledge in school settings. Regardless of how knowledge is represented by students, the teacher acknowledges that there are many ways of knowing, and that these can offer unique perspectives on the phenomenon or event being explained. Some of these ways of knowing are grounded in culture (e.g., how indigenous communities explain the relationships between humans and other living things in the natural world) and are treated as legitimate, even though they differ from common school science explanations.

Teachers *use frequent formative assessments that allow students to show what they know* (through combinations of speaking, writing, drawing, etc.). This is preferred over an exclusive focus on tests and quizzes that largely identify what students don't know. Many young learners feel enormous pressure when faced with tests, and don't get any useful information from completing them, other than being assigned a grade. Formative assessments, on the other hand, can tell students where they are currently in their understanding about science ideas, and what they might need to work on. This focused feedback benefits all learners, but in particular those who seek meaningful understanding and/or who struggle with communicating what they know.

These equity features of AST all contribute to learning, but they can be too easily overlooked or considered nonessential. We don't want to fall into the trap of seeing equity as occasional acts of care and kindness directed toward students who "need help" (a deficit point of view). Equity is instead a characteristic of the instructional environment that increases the capacity for *everyone* to participate in meaningful learning activity, and requires wide-ranging professional skills to sustain each day. Equity is a moral imperative, but it is also immensely practical. More of your students will feel that science has something to do with their lives, and that their lived experiences are valuable levers for understanding science; more of your students will understand and appreciate the aims of science lessons; fewer students will push back from the table in frustration when confronted with challenging tasks; more will believe they can do the work, or if they can't, they'll know that they're in a classroom where help and resources are always there for the asking. For all these reasons, we recommend that you treat equity as foundational to your teaching.

HOW THIS BOOK IS ORGANIZED AND HOW TO USE IT

Chapter Arrangement

We start with the first of four core sets of teaching practices (planning for engagement with big science ideas, eliciting students' ideas, supporting ongoing changes in thinking, and drawing together evidence-based explanations). Chapter 2 explores how to plan at the unit level, and in particular helps you identify what the big science ideas are that students should understand deeply. This is also where you'll find guidance about selecting an anchoring event and crafting an essential question that students will start with (they will develop their own science questions along the way).

We then shift to two chapters on classroom discourse, because managing productive forms of talk is fundamental to the remaining practices that make up AST. Chapter 3 provides a basic vocabulary for the different kinds of talk moves that can be used in conversations and how such moves can be used to serve important learning goals. These include probing, pressing, revoicing, follow-ups, wait time, and others. We illustrate how different teachers, using the same lesson plan, create widely varying opportunities for students to reason individually and together, depending upon how they deploy these talk moves. Chapter 4 lays out conditions that help all students feel able and willing to participate in classroom conversations. As you might imagine, just having a set of teacher talk moves does not, by itself, transform learning. You'll need to incorporate norms, create scaffolds for different kinds of talk, use strategies for dealing with silence, prompt reluctant students to join in conversations, and encourage students to address each other rather than always communicate "through you."

Chapter 5 lays out the second core set of teaching practices, about eliciting students' ideas and activating their prior knowledge. These practices are often used on the first day of a unit so you can find out how your students reason about the big ideas and what experiences they are bringing in from outside the classroom to make sense of the topic.

We then devote chapters 6 and 7 to the scientific activity of modeling. Modeling (together with explanation) is at the heart of knowledge building in the discipline. Research confirms that, through modeling, students can understand science concepts and learn how ideas evolve, using evidence and new information. Unfortunately, modeling is a structured activity that few teachers have ever experienced as learners. It has never been part of common practice in science classrooms. These are good reasons for you to explore both the chapter on how to help students create initial models and the follow-up chapter on revising models with evidence and ideas.

Chapters 8, 9, and 10 explore the third set of core teaching practices, designed to prompt changes in student thinking about science ideas that are embedded in the anchoring event. This is where activities like investigations, readings, and lab work are used. Although these are familiar to most teachers, the way they are used in combination with talk to shift students' ideas is not at all common in classrooms today. We devote one full chapter to each core teaching practice in this set.

In chapter 11 we introduce science argumentation. Lessons that include activities and introduce ideas often result in discussions of evidence for existing explanations, and new claims by students. We lay out a basic vocabulary for scientific argumentation and provide a number of scaffolds that you can use with students.

The final set of core teaching practices is addressed in chapter 12. Here we show how to help students work together as a scientific community to draw upon ideas and evidence from multiple lessons as they make revisions to their current explanations and models. This set of practices is typically used once in the middle of a unit and again at the end of a unit.

To help you see how the core teaching practices are used together in a unit of instruction, we call your attention to appendix B. We use two classroom examples that reappear throughout the book: the story of a fifth-grade classroom where students are investigating how a singer can break a glass with just the sound from his voice, and a tenth-grade classroom where students are investigating how the Yellowstone ecosystem could be drastically altered by the reintroduction of wolves.

The book would not be complete without chapter 13, on organizing your colleagues for change. To innovate in the classroom, you need collaborators to take risks with and play the role of critical friends. Together, you can design pedagogical experiments and test hypotheses about what kinds of teacher moves or classroom routines result in greater participation and learning by students. Rough attempts at new practices can be analyzed from multiple perspectives and reworked more successfully by teams than by individuals. In this chapter we share a framework for studying and improving practice with peers.

We finish with chapter 14 by exploring an important question: Can we be ambitious every day? Here we address frequently asked questions about which parts of AST students should experience multiple times over the course of a year, which practices should make an appearance every week in one form or another, and which should be part of teaching every day. We also describe other effective frameworks for teaching (project-based learning, teaching for social justice, integrating history and science, participating in citizen science, etc.) and share how AST tools and routines can be integrated into these models of instruction to enrich the sense-making discourse, allow more students to participate, and keep the focus on big ideas.

Tools to Get the Work Started

Embedded in each chapter are examples of tools that we and our teacher collaborators have developed over the years. By "tools" we mean materials that guide or organize different types of valuable work, either by you or your students. These are remarkably useful in helping you make the shift to AST and engage your students in more challenging activity. The tools fall roughly into four categories (see appendix C for a more complete taxonomy of these tools).

- *Planning tools* help you use curriculum and standards documents to identify the big science ideas that you and your students should be spending the most instructional time on. These tools also help you develop or select anchoring phenomena and then sequence learning opportunities for your students in ways that allow them to develop cumulative and coherent understandings of these events or processes.
- *Face-to-face tools* are used with students to represent and work on *their* current ideas. These representations are often put on the walls of the classroom so that students' current thinking can be *made public and revised* over time So, these tools don't just display thinking—they help organize and refine students' reasoning as well. Face-to-face tools include small-group models, whole-group consensus models, summary tables, routines for revising models, lists for competing hypotheses, and explanation checklists. These may be unfamiliar to you now, but we provide many examples in each chapter and describe how they are used to best effect.
- *Scaffolds for talk, listening, and writing* can take the form of sentence frames to start conversations, norms for whole-class discussion that are developed

by students, roles that students can take up in small-group activity, structured ways to get students to listen actively to one another, and guides for how to help students rehearse final explanations or arguments. Scaffolding is a term used a lot in education but is poorly understood, so we define it and provide many examples of how it is used during AST.

- *Tools for improving teaching* include guides for deciding what you and your colleagues want to change and how to get started. We also share instruments for collecting data on participation from students and displays that can help teachers see data trends in students' progress around science ideas.

Not surprisingly, all of the tools we've just listed work hand-in-hand with the theory of action that goes with AST and in particular with the four sets of core practices. We've seen these tools shape our teachers' vision of what is possible in their classrooms. We've also seen teachers adapt these tools so they function more effectively with different groups of students at varied grade levels.

FINAL THOUGHTS ABOUT RESEARCH-BASED TEACHING

Over the past decade, several authoritative documents have been produced by groups of teacher leaders, scientists, and educational researchers. These have been clear and consistent about the kinds of experiences science learners need.[9] Two products of consensus—a report called *Taking Science to School* and the *Framework for the Next Generation Science Education Standards*—sum up what students (and teachers) should be able to do, from elementary through secondary levels:

- understand, use, and interpret scientific explanations of the natural world
- generate and evaluate scientific evidence and explanations
- understand the nature and development of scientific knowledge
- participate productively in scientific practices and discourse

Most teachers read this list and imagine that they could use some help with one or more of these areas. Indeed, these student competencies are not fostered by common forms of K–12 science instruction in the United States (as underscored by the breakdown of strengths and challenges listed in table 1.1). Large-scale classroom observational studies show that many teachers create dynamic and challenging lessons, but the broad trends indicate that most

TABLE 1.1. **Strengths and challenges of American K-12 science instruction**

RELATIVE STRENGTHS OF TEACHING	SERIOUS STRUGGLES
Lessons generally well-organized	Content presented as facts, definitions, algorithms, procedures
Clear learning goals	Questioning does not stimulate thinking, discourse is teacher-centered
Frequent, high-interest activity	Few connections made between activity and big science ideas
General respect for students' ideas	Student ideas or everyday experiences not used as resources

lessons focus on activity rather than sense-making discourse.[10] In addition, pressing for "why" explanations is exceedingly rare, questioning in general is among the weakest elements of instruction, and less than a third of lessons take into account students' prior knowledge.

International tests (such as the PISA studies) show that US science students struggle with interpreting scientific evidence, identifying assumptions behind knowledge claims, using evidence, understanding the reasoning behind conclusions, and applying knowledge of science to new situations.[11] These research studies tell us that as a nation we are not meeting our teaching goals, but they don't tell us *how to change our practice* so that we can achieve desirable outcomes for all students. There is research that can help shape teaching practice, but, as we have described, the translation to practice is problematic. No single research investigation definitively tells us "This is how you should teach." For example, there is convincing evidence that having students develop coherent and accurate scientific explanations for themselves, rather than being provided with these explanations, results in deeper and more integrated understandings of the subject matter. On the other hand, there are also research studies showing that judicious use of direct instruction is valuable in helping some students develop coherent scientific explanations. It turns out that powerful teaching combines direct instruction with extended opportunities for students to make sense of explanations on their own terms.

These studies never point to a right way to teach; they can only tell us that under certain instructional conditions, certain versions of instructional practice are more influential than others for particular kinds of student participation and outcomes. This is why—even with access to research findings on strategies like group work, formative assessment, and various inquiry approaches—the professional science teacher must still exercise judgment about every dimension of his or her practice every day, and not chase after simple formulas for success offered at one-day workshops. In this book we provide you with understandable interpretations of research that can help inform your everyday decisions.

When we use our core practices and tools with teachers, some of them ask a very fair question: Isn't following research-based curricula good enough? Why would we reinvent what experts have determined works? We are very clear in our response: While there are good materials out there on the market, professionals can never place their faith in the notion that the curriculum is the one tool—the Swiss army knife of the profession—that will do all sorts of decision making for them. Only a small number of curricula that use the label "research based" have actually documented what students have learned from the activities and made changes in the materials based on those evaluations.

Even if you have an excellent curriculum, it is a mistake to imagine that it does the work of organizing how you should teach your students. What teachers do in the classroom is, in fact, more influential for students' academic futures than any other in-school factor including quality of curricula, type of school, length of school year, or peer influences.[12] Our vast national investments in new standards, curricula, assessments, technology, and professional development translate into student learning only through the many face-to-face encounters that teachers have with students every day, and the decisions made in those moments to press for or retreat from deep understanding, to engage in authentic science activity or prepare for the next high-stakes test, to encourage students' identities as capable knowers of the natural world, or to politely answer all their questions for them. That's a serious responsibility to be placed on our shoulders, but it's one that should motivate us as well.

TWO

CORE PRACTICE SET #1

Planning for Engagement with Big Science Ideas

NOT EVERY SCIENCE IDEA in your textbook or curriculum is worth teaching. Instructional materials often contain a sprawl of science ideas that don't cohere into a bigger picture for learners; they call for far more activities than teachers could reasonably do with students, and in the process, inject unnecessary technical vocabulary.[1] As a result, students can experience instruction as a series of unrelated lessons. They may not understand how readings or new concepts fit together to create usable knowledge for figuring out problems in the world. Without a focus on big ideas, the goals of hands-on work can also be puzzling for students. When asked why they are using materials and equipment, they often say, "Because the teacher wants me to," or "We do labs on Wednesdays."

There are solutions for these issues. In this chapter we'll describe a unit planning process in which you start by identifying the most important science concepts in your curriculum and standards, and leave out trivia so that your students spend more time building knowledge about powerful ideas. We'll then show you how to identify complex and puzzling events that both embody these science ideas and interest students. These events can anchor their efforts to understand core disciplinary concepts in meaningful ways.

This set of planning practices, illustrated in figure 2.1, is a systematic design process guided by clear learning objectives for students. The overarching goal is to construct or modify a series of lessons that help students build knowledge

FIGURE 2.1 **Core practices: Set 1**

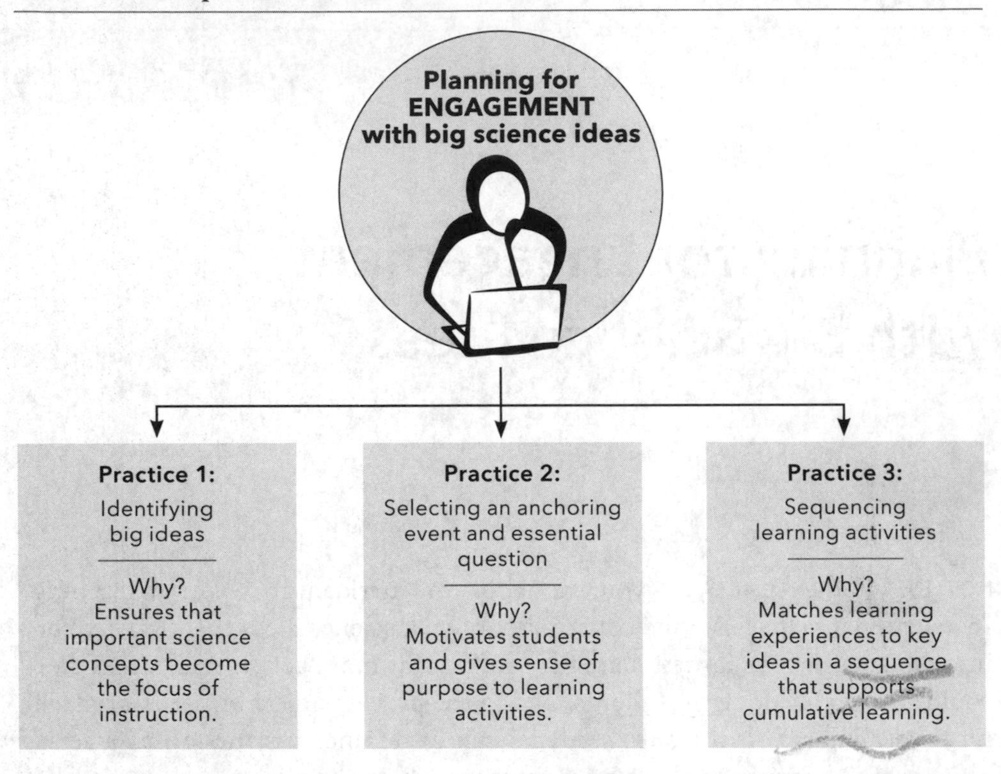

in coherent and cumulative ways. This unit planning process puts challenges on the horizon that students care about. To be clear, when we refer to a "unit" we mean between three to four weeks of instruction; this is the time scale on which teachers can most easily see changes in student understanding.

The three planning practices that make up this set are as follows:

- identifying a group of science ideas that are fundamental to explaining most phenomena related to the topic that students will be studying
- selecting an anchoring event and essential question that will support sustained intellectual challenges for students over the course of a unit
- sequencing learning activities in ways that help students build cumulative and coherent knowledge over time

PRACTICE 1: IDENTIFYING BIG IDEAS

Science topics such as earthquakes, solubility, or cellular respiration are starting places for unit design. Unpacking these topics begins with a look at the standards and at your curriculum materials. Whatever standards you use will likely have statements about what students should know and be able to do. Whatever curriculum you use will likely have key ideas printed in bold or used as chapter titles or section headings. Many teachers, however, tell us they don't have an official curriculum, or that they are expected to invent it themselves; in these cases, they look primarily in the standards for their big ideas. The core practice here is to identify the big ideas in the curriculum and/or standards that relate to the topic students will be studying.

The next move is to identify how these ideas might be related to one another and which are *most* important to teach. You'll prioritize two or three of the most powerful ideas to focus the unit on, then identify a second tier of ideas that support the development of those core ideas, and finally discard many ideas that would be nice for students to know but would likely detract from their efforts to deeply understand the concepts you've marked as most important. The routine we use to coordinate this work has been referred to as the *whiteboard activity*, and it is best done, of course, with colleagues.

The Whiteboard Activity Step-by-Step

Identifying ideas in standards and curriculum

Start by looking at the standards you are responsible for. Identify those that relate to your topic and write each on a sticky note. In figure 2.2, across the top of the whiteboard, we show four standards that a group of middle school teachers identified as being relevant to a unit on earthquakes they were about to teach. Next, take a look at your curriculum (if you have one). As you page through it, note the most prominent ideas that are mentioned. These are often printed in bold or used as headings in the teacher's guide. Select up to five of these ideas—more than five is likely too big a chunk to explore in depth—and write each on a sticky note. The bottom row in figure 2.2 shows ideas from the middle school earthquake curriculum. If you have no curriculum, you should select six or so ideas from the standards.

We've learned that teachers find it difficult to do this work unless they have "cheated a bit" and started to identify a tentative anchoring event that relates to the topic. You do not have to decide on the perfect anchoring event at this point in the process; just a rough idea is enough, and you can even keep two or three possibilities in mind if you like. Write out one sentence that describes your anchoring event(s), then get back to the sticky notes.

Clustering ideas that are related

Now start playing with different arrangements of the sticky notes—representing how the ideas are related to one another. You can do this by placing some closer to others, or you can use other strategies that make more sense to you. Most of our teachers prefer to move the more central ideas closer to the center of the whiteboard, and the tangential ideas to the periphery. Take at least twenty minutes to sort and re-sort, thinking out loud with your colleagues until your arrangement is somewhat stable. Even as you move these sticky notes around, bear in mind that ideas from the standards have to be given priority—you can't just leave them on the margins.

Identifying ideas with the power to explain

Now you'll need to determine which two or three ideas have the greatest explanatory power. To do so, ask yourselves the following: Which of these ideas, if our students could really understand them deeply, would help them explain a range of other ideas in this unit? In our example with the earthquake unit, the teachers initially put at the center of the whiteboard "Convection drives plate movement, uneven heating from the core." But it was unclear what they thought was really important about this. We urged them to write the idea out in a full sentence (this is difficult). They struggled for about ten minutes, then wrote: "Motions of the mantle and its plates occur primarily through thermal convection—the cycling of matter due to outward flow of [heat] energy from earth's interior and grav mov't [gravitational movement] of denser materials inward." This is a much clearer expression of a big idea (figure 2.2). It also fits nicely with three criteria, in the form of questions, that we developed for big ideas:

- *Can this idea help explain a range of other ideas in the unit?* Yes, in our preceding example it helps explain seafloor spreading, earthquakes, some of mountain building, and where the long-term heat source in the earth comes from.
- *Does this idea express relationships, not just facts?* Yes, it expresses links between heat and buoyancy of the mantle material, between movement of mantle materials and plate movement.

FIGURE 2.2 **Construction of "big ideas" about earthquakes from collaborative whiteboard activity**

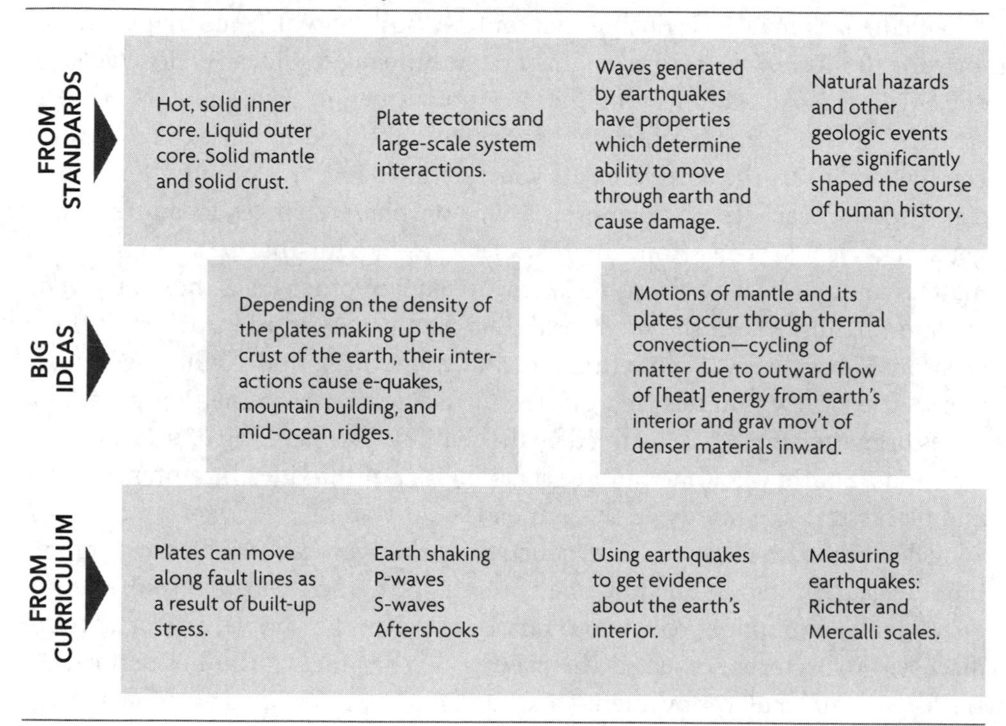

✓ *Is this idea written as a full sentence so we are clear with one another about what it entails?* Yes, the original phrase "uneven heating from the core" was unpacked in more detail.

After a few minutes, these teachers had a second big idea: "Plate tectonics can be viewed as the surface expression of mantle convection." They felt, however, that this needed more refinement because it did not meet the first of our three criteria, and it seems to just repeat the other big idea. They changed it to: "Depending on the density of the plates making up the crust of the earth, their interactions cause e-quakes, mountain building, and mid-ocean ridges." They had an "a-ha" moment, and we agreed that they had landed on the second of two powerful ideas.

Challenges That Pop Up During the Whiteboard Activity

Two challenges may arise during this routine, both of which can be generative for your thinking. One is that you may realize the biggest idea is *not actually on any of your sticky notes*, meaning that it is not named in your standards or curriculum. This happens more often than you might think. If this is the case, then you have to insert the core concepts yourselves. Here's an example. A group of eighth-grade teachers was planning a unit on phase changes in matter. They got stuck because ideas from their curriculum, like melting point or states of matter, couldn't be used in any clear way to explain other ideas, nor could they explain phenomena in the real world. But then we asked the teachers if there was some idea that was more fundamental, something underlying these other ideas. After a few minutes they started to realize that both melting point and states of matter could be explained by the making and breaking of intermolecular bonds. So, they got out another sticky note, put that idea in sentence form, and placed it in the middle of the whiteboard.

The second challenge, also productive, is that you realize the most important ideas are not found on any one note, but rather in *the relationship between the ideas on two or three notes*. In our phase change unit example, after a bit more discussion the teachers asked themselves, "What makes the bonds break?" It's kinetic molecular movement—a kind of energy associated with heat. They got another sticky note out and wrote "kinetic molecular energy," putting it right next to the "making and breaking intermolecular bonds" note. This pair of ideas, then—or, more precisely, the relationship between them—could help explain nearly everything in the unit. They later combined these two ideas into a single sentence, which helped them clarify as a group what the relationships were.

This is not a process that you should feel closure on in just an hour or so. It's messier than we've described. Most teachers work on this with peers for a while, then come back a day or two later with fresh insights they had while riding the bus, doing the dishes, or taking a walk. This is simply part of the same creative process that other professionals like artists, architects, and engineers use to break down and reassemble ideas for a purpose. One group of teachers expressed their experience this way: "It was really hard, but worth it in the end."

PRACTICE 2: SELECTING AN ANCHORING EVENT AND ESSENTIAL QUESTION

The selection of anchoring events happens in the middle of the planning process, but before we go into the guidelines for doing that, we want to introduce some background to help you understand what the larger planning process is building toward and why anchoring events provide unique opportunities for students to learn content in cumulative and compelling ways.

Why We Focus on Action in the World

The term *anchoring event* is shorthand for any complex phenomenon for which students can develop models and explanations over the course of a unit. Examples include solar eclipses, the human body maintaining its internal temperature in extreme environments, the spread of invasive species in an ecosystem, or cars crumpling in crash tests. Anchoring events can have historical significance, such as the survival of particular species of finches on the Galapagos Islands. They can be about issues of social justice, such as the rise of diabetes or asthma among adolescents who live in poverty-impacted communities. They can be about epic phenomena like tsunamis or everyday occurrences like the spread of smells through the air.

We will use terms like *events*, *processes*, and *phenomena* in this chapter just for the sake of variety, but we don't want to confuse you. Events usually take place in a short time span (solar eclipses, crash tests for cars), while processes usually take place over longer periods (repair of tissues in the body, natural selection). Events and processes both fall under the larger category of *phenomena*, or in other words, "things that happen." There are no hard-and-fast rules for when to use one term versus another.

What all good anchoring events have in common is that they motivate students to try to explain what is going on. These explanations are elaborated and evidence-based accounts—ones that require young learners to draw upon a wide range of science concepts and to engage in multiple investigations in order to construct narratives and representations for the target phenomenon. The process of scientific modeling, which we explore more deeply in upcoming chapters, is a companion disciplinary practice to explanation. Modeling has many benefits for learning, but its most powerful feature is that it makes student thinking more visible, and in different ways than the written or spoken word.

Anchoring events are precious commodities in AST networks: when teachers develop good ones, those events are adopted by others very quickly, becoming modified and improved over time. That being said, we want to make clear that anchoring events cannot be the basis for *all* learning in a unit of instruction. No event or process can embody every single science idea worth teaching. There will be days when you have to explore concepts that may be only tangentially related to the anchoring event you've chosen, but these days are the exception rather than the rule. An anchoring event and its essential question, then, are not the only things learners study during a unit, but they do represent the focus of students' ongoing attempts to build knowledge, individually and together as a class.

The Emphasis on "Why" in Explanations for Anchoring Events

In AST, developing causal explanations and models becomes a centerpiece of classroom activity. This is because constructing explanations for natural events is at the core of what scientists do. But it is also because this is the most demanding kind of intellectual work you can ask students to engage in. They have to link many science ideas together and understand how evidence might support different parts of their explanations. The development of these types of explanations and models takes many days, or even weeks, of steady construction and regular revisions.

Causal explanations and models are not assembled simply to describe in detail what happens during an event or process, but rather to account for why the phenomenon unfolded over time the way it did. The "why" in most of these explanations almost always involves things and actions that are *not directly observable*. This cast of unseen characters, things, and events influences the phenomenon (what we can observe) in some way. These components may not be directly observable for several reasons: they may work at such a small scale (atomic bonding or genetic recombination); they happen so quickly (electricity moving through a circuit) or over long time spans (evolution, glaciation, formation of solar system); they are inaccessible to direct observations or measuring instruments (events in the interior of the earth, neurons firing in the brain); they are abstract (unbalanced forces, concentration gradients, alleles); or, very likely, they are some combination of these. In upcoming chapters you'll see specific examples of how such unobservables are used by students to explore what is observable.

This simple relationship (using unobservables to explain the observable) has worked for our network of teachers at every grade level and in all science domains. It also happens to be the basis of authentic science explanations and arguments. To help you differentiate between levels of explanation, we've included appendix D to show contrasts between "what," "how," and "why" accounts by students. Some teachers write out explanations at each of these levels for the focal event of the unit, so that they can better plan for supporting students to reach the "why."

Why We Focus on Anchoring Events and Not Just Topics

Units built around anchoring events are very different from a set of lessons assembled around topics or themes. Topics are like chapter headings in textbooks—ecosystems, chemical reactions, plate tectonics, and so on. These are, of course, legitimate starting points for planning; however, when teachers use topics as the only organizing frame for unit development, they are free to put anything and everything into the curriculum. While all lessons in a unit might be nominally related to a topic (e.g., cells, or force and motion), it's usually only the teacher who sees how the activities and readings hang together. Students, on the other hand, are left to guess what all the activities, taken as a whole, are supposed to mean. The same principle applies to instruction based on themes such as "art in science" or "building a future with sustainable resources." These units are often a collection of activities that are related only in name to a bigger set of ideas. We are not against the teaching of innovative and dynamic material, but we do recommend that it is addressed in the context of explaining anchoring events or projects in which students design some authentic product. These tie lessons together and consolidate student learning of important ideas as units progress.

The research literature on learning is clear that students develop deep understandings best from being engaged with complex problems. This is not a "basics first" approach in which students learn disconnected facts, vocabulary, and formulas before doing any meaningful work. Rather, it is *within the quest to understand complex problems* that facts, terminology, concepts, mathematical equations, and lab skills become useful to students. This is true for everyone in the classroom, but in particular for English language learners, who depend upon the context provided by well-chosen phenomena in order to make sense of the words being used and the science activity as well.

Picking Back Up with the Planning Process

So far in your planning process you have selected a subset of science ideas from your curriculum and standards that are important to teach, and identified two or three of these as core to the unit you'll design. You have also moved other science ideas and vocabulary "off the table." Now, with the most important targets in mind, you will make the final decisions about a compelling science event for which your students can create evidence-based explanations and models over the course of a unit.

Anchoring events are phenomena rather than topics or questions

Anchoring events are characterized by a set of circumstances that unfold over time in particular ways. This is what makes them different from curriculum topics or themes. Is chemical change a phenomenon? No, it is more like a topic. On the other hand, your bicycle rusting in the back yard is a phenomenon—it takes place over time, and has a beginning, middle, and end. Knowing the parts of the cell and what they do is a set of facts, but the uncontrolled spread of cells within a body (cancer) is a phenomenon.

Explanations for anchoring events require students to integrate a number of important science ideas together

In a high school unit on ecosystems in Washington State, several of our teachers have focused students on the puzzle of why killer whale populations are declining in Puget Sound. There are several viable explanations for this phenomenon, but they all require that students understand the ideas shown in figure 2.3, and how they work together. We chose this example because it illustrates the principle that understanding any one important science idea requires that students see how it can be used *with* other concepts to make meaning of complex events and processes.

The integration of a number of such ideas into explanations is what makes some anchoring events more rigorous than others. This idea applies to any grade level. As we mentioned in chapter 1, one of our kindergarten teachers had her students explore the question "How can somebody little bump somebody big off the end of a playground slide?" Students' explanations and models at the end of the unit incorporated the ideas of friction, gravity, forces as pushes

FIGURE 2.3 **Ideas that students must integrate to explain the decline in orca populations**

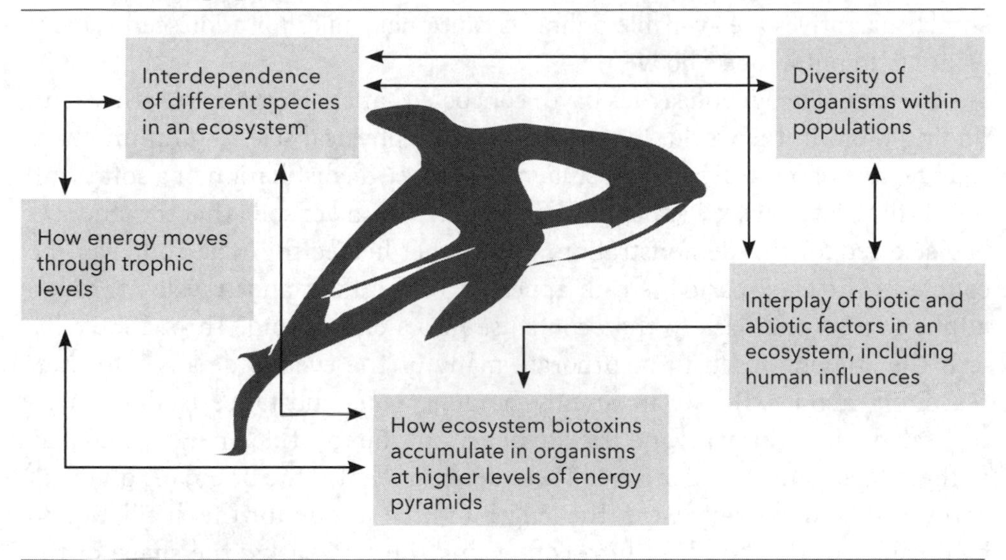

or pulls, and the relationships between a person's position on a slide and the energy released in a collision at the bottom.

Anchoring events are more compelling to students and more complex to explain if they are context-rich

Context-rich anchoring events are about a *specific* event that happens under *specific* conditions. So the question for teachers is how to transform the phenomenon from a generic science idea into a context-rich case with details. Can earthquakes, for example, be about the case of a particular pattern of earthquakes in a unique geographic region? Or perhaps a comparison between two earthquakes of equal magnitude that resulted in differing amounts of damage in two specific areas because of where the focus was located? Can cellular respiration be about a young person who has been hospitalized because a dietary supplement she uses has interfered with her ability to metabolize glucose? The details—which often include real people, places, and situations—make these stories compelling to students. The details also matter for how students put the explanations together; in the dietary supplements example, they wanted to know if the person had been taking these pills for a long time, how old she is, and if the damage is reversible. The story is true, unfortunately, and many similar narratives are available online as cautionary tales for adolescents about what *not* to put in one's body.

Occasionally our colleagues have "concocted" anchoring events rather than finding them in the world. In the eighth-grade physical science unit on phase change, one of our teachers had students observe the distillation of a soft drink on the first day. A flask was partially filled with a brand of soda that the students had selected for this demonstration. The teacher heated the soda and, over the course of the class period, steam appeared in the flask and traveled through a tube to a second flask to then condense into a clear liquid. Throughout this unit the students tried to incorporate many of the event's details into their reasoning about why an apparently homogeneous substance (soda) can be separated out in this way, and tried to figure out the role that energy could play in their explanations and models. These details included the brand of soft drink and what its ingredients were, the length of time it took for the distillation to be completed, the length of tube connecting the two flasks, the shape of the flasks, and the fact that a stopper on one of the flasks had popped off twice during the demo. These specifics are precisely what make the anchoring events interesting to students. Some of these details mattered to the explanation and some didn't, but this is another positive thing about anchoring events: students learn to discern what information is useful to their knowledge building versus what information is simply a distraction.

Good anchoring events allow students to construct different types of explanations and models

We don't want students to reproduce the explanations of others. In the soft drink distillation example, some students focused on increases or decreases in the kinetic energy of the molecules during their trip through the apparatus. Others oriented their explanations around the mutual attraction of the water molecules because of their polar structure, and others foregrounded the compounds that were dissolved in the soft drink. Students' accounts varied according to what part of the phenomenon they saw as anchoring their sense making. Students were held accountable for all aspects of the science, but they were allowed to show what they knew in ways that made sense to them. As you read upcoming chapters you'll notice this theme—students can express their understandings in many legitimate ways, and you will want to set up the conditions for them to do this.

Whenever possible, make anchoring events or situations relatable to students' interests

Tying anchoring events to students' interests increases the probability not only that students will be motivated to learn, but also that they will capitalize on their background knowledge and everyday experiences to develop explanations.[2] The most relevant context for study would be some aspect of students' lived experiences (i.e., relating to students' home, school, peer culture, or out-of-school experiences). The second most relevant context is the local community, physical surroundings, or history of a region where students live. The third most relevant context may not currently be connected to students' worlds but could be important to their lives or careers beyond school. All three levels of context are important and should be considered when you are selecting an anchoring event.

When you start working on anchoring events, you'll reach the limit of your own subject matter understanding very quickly. You should begin looking at various resources on the web or in texts to expand what you know about the topic. As professionals, we can never assume that we know enough about the subject. It is important also to work with your colleagues, asking them how they understand the explanations, models, and other ideas related to the topic in the curriculum.

Developing Essential Questions to Go with Your Anchoring Events

A good anchoring event is made even more effective when it is accompanied by a compelling essential question (also known as a *driving question*). An essential question cannot be answered with a yes/no response; it requires critical thinking. A sample essential question for a unit on cells in biology might be "What makes wounds heal in different ways?" For a unit on the respiratory system, an essential question might be "Why is asthma so prevalent in poor urban communities?" For a unit on oxidation in chemistry, an example might be "What keeps things from rusting, and why?" For a unit on forces in physical science, an example might be "How does this pulley system help me lift things that are heavier than I am?"

In a middle school unit on earth-moon-sun interactions, one of our teacher colleagues had planned for the essential question to be simply "Why do solar eclipses happen?" Her own students, however, began talking about more interesting questions, some derived from their own families' experiences. These became the focus of the unit: "Why are solar eclipses so rare and so predictable?" The students' questions are better because they'll take them beyond understanding why these events occur (which students could look up in a lot of places) to understanding how gravity has kept the earth, moon, and sun tied together in regular patterns of movement over millions of years, and the different ways that the moon's shadow can cross the earth's surface. Here's another case we briefly mentioned in chapter 1: In preparing for a high school biology unit on population dynamics in ecosystems, a group of teachers decided to use the reintroduction of wolves to Yellowstone National Park as the anchoring event. The original essential question was "How could the reintroduced wolves cause so many changes in the ecosystem?" But the teachers did not think this would compel students to include in their explanations how the social behavior of animals like wolves can result in sweeping changes to the behavior and survivability of other populations. So they suggested a small change that made a big difference in what was expected in students' explanations: "How can *so few* wolves cause so many changes in the ecosystem?"

Some essential questions, like the ecosystem example, require students to explain the anchoring event. In other cases, such as with the solar eclipse, the final pair of essential questions (why so rare and so predictable) does not specifically require students to explain why the phenomenon occurs. If this latter situation is the case in your unit, you should make it clear that students will indeed be responsible for producing explanations and models for the event, perhaps doing this "on the way" to answering the essential question(s). Essential questions, no matter how interesting, should never excuse students from developing evidence-based causal accounts for the anchoring event.

PRACTICE 3: SEQUENCING LEARNING ACTIVITIES THAT BUILD SPECIFIC UNDERSTANDINGS

In this final planning practice, you select and arrange learning activities for your students that will help them wrestle with the big ideas of your unit. You begin this by outlining an explanation for your anchoring event. Include unobservable events and processes to form the explanation. This should be *gapless*, meaning that it includes necessary details and does not leave out any links between causes and effects. It should be written as "just beyond" what students at your grade level might be capable of. You are not aiming for them to reproduce your explanation. Rather, one purpose of writing it is to prime you to be responsive to the prior knowledge and out-of-school experiences students draw upon that might be relevant to the explanation. You'll be surprised how well your preparation works when eliciting students' ideas at the beginning of the unit.

There's a second reason to write an explanation yourself. If it incorporates most of the important science ideas related to the overall topic (as it should), then you can use it to identify activities, investigations, readings, and types of conversations that should make up the unit. For example, in chapter 1 we mentioned a group of fifth graders trying to explain how a singer could break a glass with just the energy from his voice (we've had versions of this unit taught at third, fifth, and ninth grade). Figure 2.4 shows the fifth-grade teacher's explanation, with distinct science ideas indicated by a black bar in the left margin. Each of these was matched up with one or more learning experiences for students. Some of these came from the curriculum, while some had to be added from other sources.

Once you've selected the learning experiences, you can start arranging these in a sequence that makes sense for students. In our work with teachers we do not try to agonize over a perfect sequence, so we ask ourselves a simpler question: If we divided the unit into three parts, then which activities, readings, investigations, and so on, should go in the first third, middle third, and final third? In doing this "roughing out" we have three rules of thumb that are used together. The first is that ideas from the explanation that are more directly linked to students' prior knowledge or everyday experiences should go earlier in the unit. In contrast, ideas that are abstract or purely conceptual can be introduced later and build upon the more concrete ideas introduced earlier. The second rule (which can be used with or in place of the first one) recognizes that some ideas are built from other ideas—they are like composites. These we put later in the unit, and we put the more elemental ideas that make up these composites earlier in the unit. In the sound energy example, resonance is a complex idea that requires students to understand more elemental concepts—like sound as vibrations, sound as waves, and frequency as a quality of waves—so it would not be introduced early in the unit. The third rule is that most events or processes have a beginning, middle, and end, chronologically speaking (in

FIGURE 2.4 **Sequencing activities (on left) that match up with the explanation for the singer shattering the glass (on right)**

Arrows indicate only a sample of the connections.

Making and sensing sounds in the body

1. Vocal cord and eardrum readings–How do we make sound when we talk and sing? What parts of our body help us make loud sounds (vocal cords, diaphragm)? How do we hear sounds?

Travels through a medium

2. Recess ball bounce observations–Sound of bouncing ball travels in all directions, can feel vibrations in floor, echo off walls; volume decreases with distance.

Vibrations

3. Tuning fork vibration exploration–How does force relate to volume? How does size relate to pitch? How does a tuning fork make sound?

Energy

4. "Sound is energy"–The force of sound; What does "volume" mean?

Distance

5. Sound in the ground–Listening through tables, doors, and floor for sound. How does sound travel?
6. Hallway investigation–Volume of sound decreases over distance.

Resonance

7. Hum drum activity–Bowl begins to resonate by humming, vibrating bowl makes salt "dance" that is different motion than blowing salt (wind moving air vs. vibration of air).
8. Grandpa John resonating tuning forks video clip–Demonstrates how one tuning fork can vibrate a "twin" fork across a distance without touching (resonance).

Amended notes after teaching:

9. We did not plan to go into the level of molecule movement. However, students brought up the idea of particles so some small group discussions focused on the idea of particles.

Sound energy is closely related to mechanical energy. When an object vibrates, such as vocal cords, the vibrations push the surrounding molecules in every direction (4). A medium is necessary for sound energy to be moved from one place to another because the sound travels by pushing molecules into each other in all directions emanating from the source, like dominoes (2 and 5). A medium is composed of molecules and can be solid, liquid, or gas.

Each sound has properties like frequency and volume. Higher pitch sounds have higher frequency. This means that the vibration is pushing surrounding molecules in pulses at a high rate over time. If it also has high volume, then the vibration pushes the molecules harder at that rate. For low pitch, the sound wave has lower frequency. This means that the vibration is moving molecules at a slower rate (more time between pushes). Again, more volume is a push with more energy and the vibration will be larger, but the rate of vibration will stay the same if it is the same pitch (6 and 4).

The singer uses muscles to make his vocal cords vibrate (1). The vibrating vocal cords push air molecules that are all around us and inside our trachea and lungs. The push is not constant, but a pulsing compression wave (3). These vibrations travel through air. Air molecules do not move across space but rather they push against neighboring air molecules. This domino effect eventually pushes air molecules into our ears (2 and 5).

We hear these vibrations as sound. Our eardrums vibrate at the same rate as the original vibration so we can hear (and feel if loud enough) him singing the note (1). When the singer hits the particular note, the wine glass begins to vibrate, too (3). The singer listens to hear if the glass is making noise, too (resonating). You can find this note by tapping the glass or running your finger around the edge.

The closer the singer is to this pitch, the more the glass vibrates (7 and 8). Then, as the singer increases his volume on that same pitch, the wine glass really vibrates (6). The glass is flexing and wobbling. Air molecules are always bumping against the inside and outside of the glass. The glass flexes because air molecules bump into the outside of it with more force than usual (extra force comes from sound wave).

The glass will stay together through a little bit of wobbling, but over time the flexing back and forth will break the glass (9). So the shape of the glass is important. The wine glass is thin and round. The tiny stem does not absorb a lot of the vibrations. A thicker glass could absorb the sound vibrations and would not wobble or flex as much, so it wouldn't break (2 and 5). The glass also has to be empty (9). The empty glass will flex and wobble at a certain pitch (7 and 8). However, if there were water in the glass, the water would help the glass keep its shape. The water would push on the sides of the glass and keep them from wobbling or breaking. The sound would travel through the water, too (9).

contrast, some are cyclic or are multiple events happening simultaneously), so it is helpful for students to understand the science behind the beginning stages of an event or process first, then explore other ideas later. Now that we've said there are three rules, we should probably add a fourth: none of the first three rules takes precedence over the others. You just have to use your judgment about when to apply them.

The left side of figure 2.4 shows the list of activities and readings for the shattering glass and how they are matched up with the elements of the full explanation. The teacher used a combination of the first and second rules we just outlined. She wanted to start with activities that were close to students' personal experiences, so, for example, she put a vocal cord/eardrum lesson first. In the middle of the unit students applied ideas about vibration to new situations—tuning forks, sound energy dissipating over distance, and using different media to transmit sound. In the last part of the unit students were asked to use these ideas, together, to understand the complex idea of resonance.

We should add here that by the end of these units, students are expected to do more than construct a linear account of why the glass shattered. In AST, students are helped to understand, in some depth, what each part of their explanations or models means. Students should be able to *expand their explanations* in different places. For example, they should be able to expand on the nature of energy transfer by responding to the questions: Does all the energy end up in the glass? Why or why not? Students should be able to expand the part of the explanation about hearing in humans, describing how people can tell what types of sound are being made by vibrating objects. In the fifth-grade version of this unit, students should be able to expand on the nature of compression waves and sound traveling though media, by explaining how a sound wave is similar or different from a surface wave in a pond. In our ninth-grade version, students have to expand upon why a logarithmic scale is needed to measure sound energy in decibels. Expandable explanations are the result of going in depth on key ideas in your unit, and being purposeful about linking them with other ideas.

Tightening the Triangle

As your planning moves into its final stages, you can do a check on how connected learning activities are with the gapless explanation and the standards that apply to your topic. We show these connections using a triangle, with the vertices each representing planning choices you make. Our teachers have developed a set of questions they ask themselves near the end of the design process (see figure 2.5); this check is referred to as "tightening the triangle." Why do we do this? Back in chapter 1 we listed the seven elements of rigorous and equitable teaching, the first of which is "Teachers anchor students' learning experiences in a complex and puzzling science phenomena." Now we've planned that part. But using an anchoring event is only partially effective for learning; students also have to build ideas about these phenomena cumulatively, over time (another element in our list). This requires close connections between the anchoring event and the sequence of learning activities—connections that become "visible" only when the newly designed unit is examined as a whole. Similarly, it is best to check at this stage if you have plans for helping students use ensembles of scientific practices to make sense of the phenomena (another

FIGURE 2.5 **Questions about connections between three parts of the planning triangle**

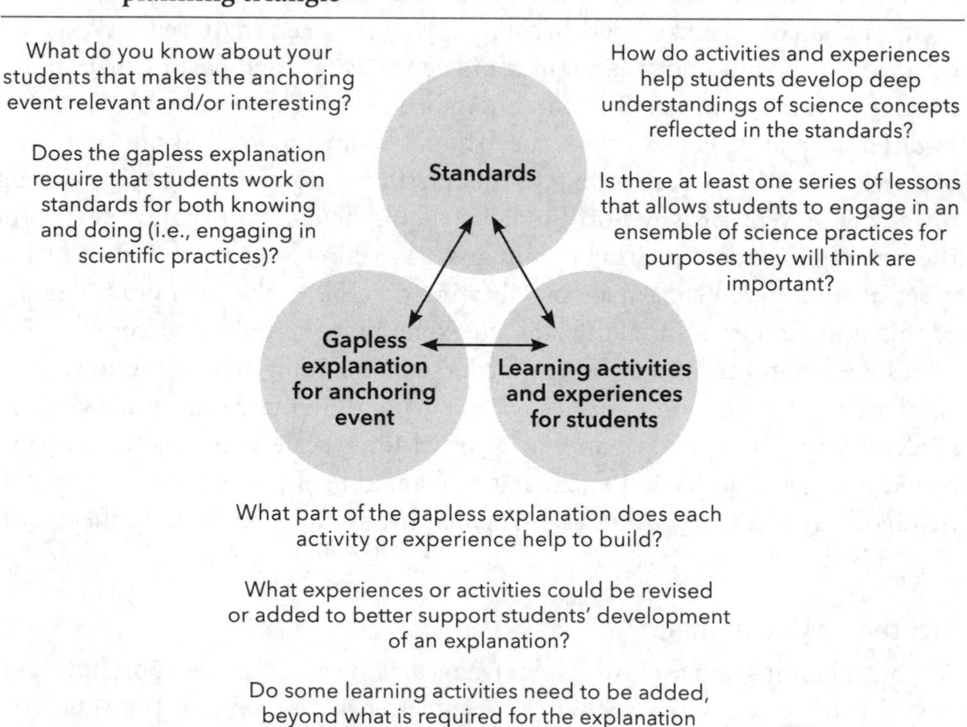

of our elements). Because many state and district standards now include student performances like making arguments, carrying out investigations, modeling, analyzing data, and communicating information, it is important to see where you've built in opportunities for students to engage in connected episodes of authentic science work and how these activities build on one another to help them develop new knowledge.

Sometimes disconnects in the triangle are not evident until you teach the lessons, and that can't be helped, but you do want to avoid major issues up front, such as not providing learning experiences that match up with important standards or having students do activities that they don't perceive as helpful in understanding the anchoring event. Keep the triangle tight, and you'll be able to support learning that students recognize as moving them forward on big ideas.

HOW TO GET STARTED WITH COLLEAGUES

We recommend that you and your colleagues devote several meetings during the school year to designing your first unit, or identify time during the summer for this work. If your school allows science teachers to determine how regular professional learning time is used, then you could petition your administrator with a statement of rationale for why this planning project is worthwhile. If you have a district instructional coach, enlist that person to help you. The planning process itself happens in cycles. Each cycle includes setting achievable short-term goals, working together on one of the planning phases described in this chapter, going off and thinking individually for a few days, and then coming back together to compare ideas and make decisions. A total of five or six meetings is a reasonable estimate to develop a unit, and this takes into account the time in between for each of you to learn more about the science content.

You can elect a unit to adapt that you have taught before, one that takes about three to four weeks. Don't feel like you have to reinvent the wheel by creating all new learning activities. Budget a couple of meetings for each of the core planning practices (identifying big ideas, selecting an anchoring event and essential question, and sequencing learning activities). Appoint one member of your group to keep track of documents and other resources that you identify or develop by making them available online. Post pictures of planning products like the results of your whiteboard activity, links to videos that represent potential anchoring events, and explanations for your anchoring event that can be edited by the group.

In the planning process we often see teacher groups stall out over selecting the perfect anchoring event. It's fine to have one that is just "okay." Attend more to identifying the big science ideas and developing a sequence of activities that students will use to create explanations and models over time. One by-product of this planning process is that everyone can see where they need to learn more about the science itself before teaching about complex and contextualized phenomena. For this reason, it helps to establish norms around talking candidly about the gaps you all have in your content knowledge.

Perhaps the best piece of advice we can give is to simply get started. You'll learn best from doing the work, and it gets easier each time you do it. Keep in mind, however, that you cannot convert all your units in a single year. Developing two or three really well-designed units over the course of a year is excellent progress.

THREE

PRODUCTIVE DISCOURSE, PART 1

Talk as a Tool for Learning

MOST STUDENTS ALREADY EXCEL at talk. They can exchange words with others to express how they are feeling at the moment, to get questions answered, or to make plans for later in the day. But unlike conversations in their everyday life, discourse in science classrooms can feel strange. There, students are asked to *compare* different science explanations, *argue* with evidence, *critique* a model, or *unpack* a claim. They are expected to take up technical vocabulary and to follow rules for participation that aren't always clear. Students wonder: *What am I supposed to say? Am I using this word correctly? Is it okay to disagree or ask a question?* They can feel stuck in the middle of unnatural discussions, and the situation can be especially awkward for learners from nondominant backgrounds or those who speak English as a second language. For these reasons, teachers who want to cultivate productive academic discourse—with *all* their students—have to design regular opportunities for them to try out new ways with words, and support them throughout the process of learning to "talk science."

Fostering students' discourse is important because goal-directed talk is a chance to think, and thinking is required for higher-order learning. This may seem simplistic, but here's the logic behind it. Taking a turn within a conversation requires that you activate prior knowledge about what's being said by others, organize possible responses that will fit the flow of the dialogue as well as the nature of your audience, and then verbalize your ideas while monitoring and adjusting in real time what you are saying. This stimulates learning because translating ideas into words is not simply the "reporting out" of what is fully formed in one's head. Under the right circumstances it involves reasoning processes that give structure to loosely formed concepts and makes gaps in logic more evident for those doing the talking. In addition, if we consider learning to be not just a personal act, but also a collective endeavor in which students reason together, then talk is important because it transforms an individual's ideas, puzzlements, conjectures, and questions into public resources for others to think with. In this view, students' outside-of-school experiences, intuitive explanations, and even openly stated points of confusion can be productive for advancing a group's state of understanding. For teachers, orchestrating these types of knowledge-building opportunities requires an uncommon skill set, but one that can be developed over time.

In this chapter and the next, we will share strategies and tools to help you and your students engage in productive discourse about science. The first of these is *prethinking the goals for different kinds of classroom conversations*. Every conversation should have a purpose that students can understand, and if you define this for yourself ahead of time, then it becomes easier to plan how to initiate the talk with students, select tools and routines that can support the goals, and manage specific kinds of participation along the way.

The second idea deals with the *cognitive demand* of questions and tasks that set the stage for classroom talk. Some questions, for example, require limited intellectual contributions by students ("What are the planets in our solar system?") while others can be more challenging ("How do Newton's Laws of Gravitation help us explain the movements of the earth, moon, and sun?"). Having the right mix of high- and low-cognitive-demand questions and activities helps encourage rigorous talk, rather than just more talk.

The third idea is about *developing a repertoire of talk moves to serve the goals of a conversation*. It is much easier to manage conversations if you think of them as activities involving specialized turns of talk that are used to prompt thinking or action by others. Moves like pressing, follow-ups, revoicing, and focusing (all explained later) can be used to help young learners reveal more of their thought processes, but also help them interact with one another's ideas and reason scientifically together.

The next chapter on discourse—just to give you a preview—explores ways to support students so they feel capable of participating in science talk and become more willing to share ideas with others. We refer to this as "laying the groundwork" for discourse by establishing *norms* that students follow for engaging in civil and productive talk, by developing *routines* that students will recognize as launch points for particular types of conversations, and by *scaffolding* student talk in whole-group or small-group settings.

HOW TALK SHAPES OPPORTUNITIES TO THINK: VISITS TO THREE CLASSROOMS

Opportunities for students to reason in classrooms vary directly with their opportunities to participate in productive talk. We'll illustrate this by showing three episodes of talk, each in a different seventh-grade classroom. We refer to the teachers generically as A, B, and C. It is worth noting that we actually observed these teachers, with B and C doing their lessons in the same school, using the same lesson plan. All three teachers were starting a unit on the earth-moon-sun system, and all three began by sharing a dramatic video of a total solar eclipse shot in India. It opens with a panorama of thousands of people gathered on the banks of a river on a late summer afternoon. They turn their attention to the sky, and watch as the moon passes slowly in front of the sun. Within a few minutes, the sun disappears entirely behind the moon and the crowd falls silent as the landscape momentarily becomes as dark as night.

In this transcript, Teacher A has just finished the video and her students are eager to share their ideas:

1 *TEACHER A*: Okay, so we've just seen a solar eclipse. That's pretty cool,
2 isn't it?
3 *STUDENTS*: [several together] Yeah!
4 *TEACHER A*: This kind of event has to have three bodies in our solar system
5 interacting; can anyone tell me what they are?
6 *MICHAEL*: The sun and the moon is what you need . . . then . . . I
7 don't know.
8 *TEACHER A*: That's good, Michael, but there is a third body in our solar
9 system involved; does anyone else want to try?
10 *SARAI*: The sun's light?
11 *TEACHER A*: Well, that is produced by the sun, but thank you for your idea
12 . . . anyone else?

13 *KATIE*: The earth?
14 *TEACHER A*: You got it! The earth is involved as well as the sun and moon.
15 Now can anyone tell us what the moon is doing to the earth during a
16 solar eclipse? Cooper?
17 *COOPER*: It's blocking it.
18 *TEACHER A*: I think you are close. Anyone?
19 *ALYSSA*: Is it, like, making clouds, or a shadow?
20 *TEACHER A*: Exactly—the moon is casting a shadow on us, here on earth.
21 *COOPER*: Is that different, like, from a lunar eclipse?
22 *TEACHER A*: Yes, it's different because in the lunar eclipse the earth is
23 getting between the sun and moon, and the earth is casting a shadow on
24 the moon. [The teacher then transitions out of the whole-class talk and
25 into an activity.]

In this example Teacher A is using discourse moves; however, these don't help reveal much about students' thinking, nor do they help students reason about what they know. Teacher A uses a common pattern of talk referred to as *I-R-E*. This stands for *initiation-response-evaluation*. The *initiation* is typically a question by the teacher that has a known, "correct" response (for example, lines 4 and 5), and requires only recall of facts or a simple calculation on the part of the student. The *response* is a word or phrase, usually offered by the first person to raise his or her hand (lines 6 and 7, 10, 13). The *evaluation* (by the teacher) is a comment signaling that the student is either right or wrong (lines 8 and 9, "That's good, but . . ."; lines 11 and 12, "Well, that's produced by the sun . . ."; and line 14, "You got it!"). Because this talk focuses on lower levels of thinking, the conversation drastically undershoots what young learners are capable of. In fact, it is not a conversation at all.

There are many variations of I-R-E (a.k.a. the "fill in the blank" or "guess what's in my head" routine), but in each case, correct answers are valued over thinking. In these classrooms, students may give partial responses and the teacher may have others fill in what is missing from the initial contribution. Some teachers are more polite than others about incorrect answers, while some are more terse. Regardless of the small differences, this sort of dialogue is problematic because it can become a running quiz show that puts students on edge. It is particularly daunting for learners from nondominant cultural and linguistic backgrounds, while favoring those who think quickly and have ready access to science vocabulary.

Teachers using I-R-E as their standard mode of interaction often cherry-pick the right answers from a few eager students and then assume everyone who is paying attention now shares a common understanding of how all the responses fit together to make a sensible whole. The teacher moves forward with the lesson under this assumption. This kind of dialogue is so common that it's been labeled as the default pattern of talk in schools. We have seen entire fifty-minute class periods in which students have endured one I-R-E sequence after another. Sometimes students are asked to respond to I-R-E questions in unison—giving a choral response—but this routine also has questionable learning value.

Thankfully, there are alternatives to I-R-E. Let's visit Teacher B now, to see what might be different in his classroom. In the following passage he has just finished showing the video:

1 *TEACHER B*: Okay, I can tell that you want to talk about what was happening,
2 but can everyone first take a few minutes to turn to a partner and discuss
3 what you observed?
4 [Partner conversations begin. After five minutes, the teacher calls
5 students back.]
6 *TEACHER B*: Who'd like to start?
7 *JADEN*: We said the sun disappeared!
8 *MARIA*: No, the clouds were coming between the earth and the sun; I
9 saw that.
10 *TEACHER B*: I'll write your observations; you keep going [writes on
11 whiteboard as students talk].
12 *CARMEN*: It's the moon we said—we saw it, it was coming closer to the
13 earth and when it gets that close we get the eclipse. It can happen
14 once a month.
15 *TEACHER B*: Okay, good. Giovanni?
16 *GIOVANNI*: The moon is putting a shadow on the sun; the earth and sun
17 are lined up, but it doesn't happen very often.
18 *TEACHER B*: Thanks, Giovanni, that's an interesting theory. Anybody
19 want to add?

20 *HELENA*: So, what Carmen said, it—the moon—has phases every month;
21 the moon can have eclipses too.
22 *GIOVANNI*: If it was clouds all the time . . . I think it is the moon getting
23 right between the earth and the sun.
24 *TEACHER B*: Okay, these are all really fascinating ideas; thanks to everyone
25 for sharing. We are going to find out the answers to these questions over
26 the next few days. [The teacher then transitions to an activity.]

This classroom conversation seems different from the previous example. Teacher B starts off by giving students time to talk with a partner. When they come back to the whole-class conversation, students offer a number of ideas: Jaden says the sun "disappeared" (line 7); Maria believes the clouds are coming between the earth and sun (lines 8 and 9); Carmen thinks the moon is getting closer to the earth and that this happens on a regular basis (lines 12–14); Giovanni claims the moon is putting a shadow on the earth and adds how rare this is (lines 16 and 17). Several students have shared their ideas, which each seemed to be a partial explanation for what they saw in the video.

Teacher B does not evaluate these contributions, nor is he trying to get students to say anything in particular, so this is not a case of I-R-E. As a result, the students' responses are longer and many of them are like mini-explanations. Teacher B, however, *does not do anything with* these comments, other than acknowledge them and record them on the whiteboard. This kind of talk is like a "popcorn conversation" in which ideas burst out all over the room, but there are no discourse moves by either the teacher or the students themselves that link these ideas together or compare or contrast them with one another. Each idea, wonderful as it is, arises and fades in a moment. The teacher does ask students to add to Giovanni's comment (lines 18 and 19), but that is the only time he interacts with their ideas. Teacher B's discourse is an improvement over Teacher A's in that students are allowed to theorize, and much of their thinking is made available to their peers. Teacher B also begins to get a picture of their current understandings—but there is much more that can be done with talk.

What might it look like for a teacher to use a broader repertoire of discourse moves and be more responsive to students' ideas in this classroom context? We'll move now to Teacher C, who finishes the video and asks students to record their observations silently for a few minutes. As they work, she walks between the desks, occasionally jotting down what certain students write in their notebooks. She then gives them time to talk as partners for a few minutes before continuing:

1 *TEACHER C*: Okay, we've had a chance to think. Can we share some
2 observations?
3 *CALLIE*: Well, I saw the sun; it goes dark, little by little, like . . .
4 *TEACHER C*: Can you say more about that?
5 *CALLIE*: I saw the moon, like, it was taking a bite out of it. The sun was
6 moving behind the moon really slow, well . . . one of them was moving
7 . . . but I was not sure which one.
8 *TEACHER C*: Does anyone want to add to what Callie has said about how
9 the sun and the moon were moving?
10 *EVA*: The sun, the sun looked like how the moon looks when it has its
11 phases—a half moon and quarter moon, like a circle is cut out of it.
12 I don't know if the sun or the moon is moving, probably the moon
13 because it is smaller.
14 *DARIUS*: It was happening fast. It was getting blocked and went through
15 phases in just a few minutes, so it, both of them, might be moving past
16 each other . . .
17 *TEACHER C:* So we have a question about whether the sun was moving, the
18 moon was moving, or maybe both, and then how fast. Okay . . . Lina, I
19 saw that you had written something interesting that you observed. Can
20 you share that?
21 *LINA*: But it wasn't really about the eclipse; it was about something else.
22 *TEACHER C*: But you were observant and your note might give us a clue
23 about what was happening. Are you sure you don't want to offer that?
24 *LINA*: Well, the people were waiting by the river.
25 *TEACHER C*: That totally counts as an observation. Does that mean
26 something to you if people are waiting for something to happen?
27 [silence for 20 seconds]
28 *TEACHER C*: Let's turn and talk to a neighbor about this question. Does it
29 mean anything if people had gathered by the river ahead of time?
30 *[STUDENTS CONVERSE IN PAIRS FOR TWO MINUTES.]*
31 *TEACHER C*: Okay, did any group have something they wanted to say
32 about this?

33 *MARCUS*: We said that if they were waiting, they knew what—it—
34 was coming.
35 *TEACHER C*: What do you mean they knew it was coming?
36 *MARCUS*: They knew it ahead of time; they were all dressed, it was kinda
37 like a ceremony, they were all looking up at—
38 *POLLY*: [interrupts] Someone predicted it. They, I don't know, but it could
39 have been on the news ahead of time; otherwise, they would not go out
40 there. It was something special? I don't know.
41 *EPHRAIM*: It was; I know my uncle saw a total eclipse in Turkey when he
42 was a boy. They all got together in their town and watched it together.
43 He's never seen one since.
44 *TEACHER C*: Did he tell you anything about how it happened?
45 *EPHRAIM*: It was just, like, people cheered when it got dark; they took
46 pictures. Then it got light again like nothing happened.
47 *TEACHER C*: [to whole class] Does anyone who has not had a chance to talk
48 yet want to add to any of your classmates' questions or ask them about
49 their ideas?
50 *BRIAN*: I have a question for Ephraim. Did the solar eclipse happen again
51 for anybody you know?
52 *EPHRAIM*: No, mmm, my uncle has the only story.
53 *LINA*: So it can't be the moon going in front of the sun.
54 *TEACHER C*: Why do you say that, Lina?
55 *LINA*: Because the moon always goes around the earth; I know it goes
56 around every month. So it—if it is the moon, then solar eclipses would
57 happen all the time.
58 *TEACHER C*: So I'm hearing three puzzles that you have all brought up.
59 The first is: Does the moon have anything to do with the solar eclipse?
60 The second is: How do people know it is going to happen? And the third
61 is: Why is a solar eclipse a really rare occurrence? Let's write these up
62 here [records their questions on a corner of the whiteboard] and we will
63 spend the next few days figuring out how we can answer these. [The
64 teacher then transitions to an activity.]

This conversation in Teacher C's classroom likely seems different to you from the previous two. Yes, it is a longer conversation than in the other classrooms, but this is because there is much more student thinking going on—more students are contributing and many are taking long turns at talk. They are not just theorizing about what is happening, but in some cases saying *why* they believe what they do (see Eva and Darius, lines 10–16; Marcus and Polly, lines 36–40; and Lina, lines 50–57). The teacher is making many small talk moves, but each has an impact on what students say and, in some clear cases, has an influence on their thinking. In line 4 she asks Callie "Can you say more?" to expand on her initial contribution. This follow-up question-asking strategy is used again with Marcus (line 35, "What do you mean?"), which begins a cascade of theorizing, questioning, and storytelling by five different students (lines 36–57).

At two different points in the conversation, Teacher C communicates the value of students' ideas by asking if anyone wants to comment on the thoughts or experiences of their peers (lines 8 and 9, lines 47–49). The latter example, about the people in the video knowing that the eclipse was coming, was initiated because the teacher had taken a few notes while circulating as students were writing individually about their initial observations. She thought then that Lina's observation about the people anticipating the eclipse could potentially provoke further reasoning by others, which it did. By the end of class, the teacher recaps the major themes as questions that will be pursued throughout the unit (lines 58–64). This is another signal to students that their ideas matter.

When we look back at Teachers A, B, and C, we can say that Teacher C's orchestration of discourse is more *responsive* to students' ideas and experiences, and that this was made possible through a repertoire of moves, adding up to more than just the sum of their parts. Not all student contributions to this conversation emerged spontaneously. At one point (line 27) Teacher C patiently waits for someone to say something and then decides to use a turn-and-talk (lines 28–30) so everyone has a chance to interact before whole-class dialogue continues.

Any of these three teachers could have explained to students what the "answer" was about the solar eclipse, but this would not provide any opportunities for students to think, resulting in only superficial changes to their mental architecture. Students would dutifully reproduce what the teacher told them on the next test, but without understanding what they were writing. Even for students who could comprehend a textbook explanation, as provided by the teacher, their understanding would be fragile and short-lived. This isn't just our opinion—this is what the research on learning tells us.

So, talk matters in classrooms. You can have a big impact on your students and their learning by encouraging productive discourse. We'll now address three big ideas that can help you foster regular talk opportunities for your students.

IDENTIFYING GOALS FOR DIFFERENT TYPES OF CONVERSATION

One reason teachers can flounder in the middle of classroom conversations is that they haven't imagined what the specific goals of the talk are. It is vital to prethink where you'd like to *end up* at the finish of a conversation, because this helps you imagine the types of talk you want to encourage from students, and allows you to select routines and tools to support this talk. By "end up" we don't mean that you would steer students toward making particular comments, or achieve some false consensus about an idea, but rather that you'd identify ahead of time the *kinds of intellectual work* you want students to do. For example, do you want students to develop a list of initial hypotheses about how a scientific phenomenon occurs (as Teacher C did)? Do you want them to make sense of an activity, and in what way? Do you want them to critique an explanation? These considerations not only help you initiate conversations, they also prime you to recognize certain types of contributions from students that, in turn, prepare you to respond without having to improvise every word you say.

Next we describe four possible high-level goals for talk that can help organize your thinking, each of which uses a different setup and mix of discourse moves (described later) to accomplish its purpose. In each example we include the idea of *framing*, a kind of initial orientation the teacher provides that invites students into the dialogue by suggesting how they can contribute and why the science ideas are important.[1] We are offering only an overview of each kind of conversation here—we'll go into greater depth in upcoming chapters.

Activating and Eliciting Students' Ideas About a Science Phenomenon

The goal of this conversation is to draw out students' beginning understandings of an event or process in the world. These kinds of conversations happen most often, but not exclusively, at the beginning of a unit. The setting may be a whole-class discussion after you show a video or image, do a demo, or relate a puzzling story to students. Here's an example of how you might frame a conversation about homeostasis—in this case, how our body responds to the outside environment to keep its own core temperature within a livable range:

> This unit will help us understand how organisms—including human beings—can cope with changes in the environment. The ideas we develop here will help us explain how our bodies deal with things like temperature extremes, low blood sugar, or a lack of oxygen. Our task today comes in two parts. First I'd like us to create a list of observations about a video showing two runners who are competing in a long-distance race on a hot and muggy day. The second task is for you to share ideas that help us begin to figure out what is happening to these runners. What you share can be something you've seen or experienced yourself, something you've learned previously, an analogy you think describes what you are seeing, or a question you have.

Following this framing, you would share the story and the video of the runners, one of whom starts to exhibit symptoms of heat stroke while the other does not. The questions you might start with are "What do you see happening here? How do you think this happened? What might be going on inside the runners' bodies that we can't see?" The setting for these conversations can shift to small groups or even pairs to give everyone more chances at talk, but at the end of such sessions, all students come back together to decide, "What do we think we know about this phenomenon? Why? What are we unsure of?" Structuring the talk this way sets students' expectations, helps them understand how to participate, and helps you monitor the progress of the discussion as time goes by.

Helping Students Make Sense of New Observations, Information, or Data

The goal of this kind of conversation is to help students recognize patterns in data or observations, critique the quality of data, or propose why these patterns exist. This kind of talk would happen frequently throughout a unit and would accompany nearly every kind of science activity that students do. The setting for these conversations is usually small group work as students engage in an investigation or look at secondhand data. The teacher in the homeostasis example might, in a later lesson, show data about how a person's body temperature normally fluctuates in cycles over the course of a twenty-four-hour period. The framing questions could be: "Understanding extreme events, like heat stroke, will be challenging, but we can start building an explanation by looking at how our bodies work in everyday circumstances. So, how can we tell a story about this person with the data that we are analyzing today? What's happening and why?"

During the small-group conversations, the teacher circulates around the room, spending a few minutes at each table listening closely to what's being said before asking questions like "What patterns do you see? What is going on at the unobservable level that might explain our observations?" or "What if we changed part of the system we are investigating; what might be the result?" Sometimes students will need more basic help by having the teacher point out specific parts of the data to focus on—"Let's just look at this part of the graph."

Connecting Activities with Big Scientific Ideas

The goal of this talk is to apply what is learned during activities or readings to a phenomenon that students are trying to explain. This conversation is different from the previous one in that students are not trying to explain the outcome of an activity or reasons for data patterns, but rather to relate what they learned during an activity to a larger puzzling science event. The setting for this sense-making talk is often whole-class conversation because you want students to hear each other's thinking. A typical question for the homeostasis example might be "Okay, now that we've made sense of the activity and our reading about how the hypothalamus detects blood temperature, can we use these ideas to figure out a bit more about our two runners?" Because students are frequently reluctant to offer responses right away, a good move is to ask them to "turn and talk to your neighbor for two minutes." As students later return to the whole-class conversation, you can prompt them to comment on each other's ideas. The culmination of this conversation is negotiating with students two or three summary statements they believe will link the activity and reading to the bigger science puzzle anchoring the unit. You might say: "Okay, what can we agree on? Where do we disagree? And what do we need to learn more about?"

Pressing Students for Evidence-Based Explanations

This kind of discourse usually happens once in the middle of a unit and again near the end of a unit. The goal is to assist students in using multiple forms of evidence, gathered over several lessons, to construct an updated explanation for the phenomenon that has been the focus of their work together. A framing prompt here might be "You've done a number of activities and readings in this unit. Now it's time to use these together to revise our explanations or models for the puzzling phenomenon."

This can be an unwieldy conversation to start as a whole class, so much of this talk happens in small groups. The teacher circulates, making comments to students about gaps in their current explanations, such as "Hmm . . . you have the beginning of the story and the end of the story, but what happens in between?" The teacher probes and prompts for the use of evidence—"Why do you think it happens this way?" The students here are typically busy revising a previous explanation or model, so the teacher can point to these and offer prompts like "Consider writing more here," or "Think about adding to your model what you learned from [activity X]."

These four types of conversations are not the only ones that you can use in a science classroom, but they are good examples of situations in which students are doing serious intellectual work. The larger point is that, if you are clear about the kinds of talk you want to foster—even by making up your own names for these different conversations—it becomes easier for you to develop rigorous questions and prompts to anticipate student contributions, respond to them, and help more students become involved in the talk.

HOW THE COGNITIVE DEMAND OF QUESTIONS AND TASKS INFLUENCES STUDENTS' TALK

Lower Cognitive Demand

Discourse starts with questions, prompts, or tasks that set the stage for the rigor of the ensuing conversations. These can roughly be characterized as having lower cognitive demand or higher cognitive demand (yes, there is a middling range of cognitive demand, but we are trying to show a clear contrast in the kinds of work you can ask students to do). Low-cognitive-demand tasks typically focus on memorization (recall), on vocabulary-level understanding (listing or describing things), or on procedural tasks that require students to follow prescribed steps or plug numbers into formulae. Many low-cognitive-demand questions and tasks have a "right answer" that can be expressed in just a few words. For example:

- Which organs of the digestive system help break down food chemically?

- Name the parts of the water cycle (or parts of an atom, a plant, or the electromagnetic spectrum).
- Given the mass and volume of these objects, calculate their density.
- How do we define the term "niche" in an ecosystem?
- Describe what our text says is the difference between solvents and solutes.

When students respond to these questions or prompts, whether correctly or incorrectly, it does not tell you much about their thinking. In fact, after they give their brief responses, you will be on the hook for asking another question, and then another (remember I-R-E?). Some low-cognitive-demand tasks can be challenging for students, but they don't involve useful learning (for example, memorizing the entire periodic table). These tasks produce the illusion of rigor, but they don't ask students to do anything *with* ideas.

There is nothing inherently wrong about some low-cognitive-demand questions or tasks. Occasionally you do need to check students' basic understandings, develop their fluency with certain types of calculations, or ask them to learn procedural skills. And it is possible to use follow-ups to these questions to expand the talk beyond recall—for example, asking students to define "niche" in the context of ecosystems, then asking if two organisms can share identical niches in the same habitat. The follow-up requires students to do a thought experiment and justify if they think the imagined situation is likely in the real world. If, however, a steady drumbeat of unelaborated low-cognitive-demand questions dominates the talk in a classroom, students will rise only to the level of what they are being asked, without using basic facts or procedures to work on more authentic tasks. You will end up training students to give short responses, and to make sure their responses conform to what some authoritative source (the textbook, a website, the poster on the wall, you) says is correct.

Higher Cognitive Demand

In contrast to lower-level classroom tasks, higher-cognitive-demand questions or prompts require that students *do something with* ideas (this is what defines reasoning). These demand more intellectual work and may not have discrete answers, which is why they are often referred to as "authentic questions or tasks." They are much like what professionals deal with in everyday life. You can, for example, ask students to unpack an idea in their own words, give an example of some science principle, compare or contrast ideas, solve nonroutine problems, justify an explanation, or use evidence to support a claim—these are just some of many possibilities.

Figure 3.1 gives example prompts that reflect high cognitive demand for learners. These are specific to the disciplinary content of science and can be adapted to be questions or tasks. For just about any lesson you teach, there should be an opportunity to make one of these the subject of classroom conversation. These are just examples; there are many more. Of course, you can't give students a steady diet of these challenging questions or tasks, either—they need a careful mix of lower- and higher-demand work to do. Sometimes you'll need to ask lower-demand questions to make sure everyone has a prerequisite understanding of a term, or recalls the lab results from the previous day, before wrestling with bigger questions. Our point is that prompts of higher cognitive demand can catalyze more reasoning through talk in your classroom than can those of lower cognitive demand.

FIGURE 3.1 **Example prompts that reflect high cognitive demand**

Compare or contrast two ideas	**Generate** a list of plausible hypotheses	**Represent** an idea in a different way
Predict outcomes based on data or a model	**Explain** why a process or event happens the way it does	**Modify** a model or idea based on new evidence or new science ideas
Infer what is happening in a system based on data	**Design** a study that answers a particular question	**Evaluate** the relevance and quality of data from a study
Justify claims with evidence	**Identify** gaps or inconsistencies in an argument	**Apply** an idea or principle or model to a new situation

DISCOURSE MOVES—YOUR TOOLKIT

By "discourse moves" or "talk moves," we mean the specific turns of talk that you might use inside of a conversation to facilitate the discussion of ideas. Discourse moves are like specialized tools in a toolkit: each category serves a specific purpose, as illustrated by the taxonomy in figure 3.2. It is important to realize that they have to be *used in combination with one another to achieve a conversational goal*. These combinations can be used to elicit student reasoning, model out loud how one thinks about a problem, encourage all students to participate, emphasize key ideas, and ultimately help students take up scientific discourse themselves. The examples in figure 3.2 are called "moves" because they are nearly always *responses* to things that students have said, rather than part of a recitation you give. These can be used at any time, such as in five-minute warm-ups at the start of class, in whole-class conversations, or when you talk with students in small groups. You should also encourage students to use these moves with each other.

The categories of talk moves include probing, follow-ups, pressing, opening up cross-talk between students, revoicing, wait time, focusing, and putting an idea on hold. It helps the conversation, of course, if you can pose good initial questions—ones that invite a range of responses.

Probing

Probing questions or prompts ask students to make public their observations, thinking, or past experiences. Usually these are preceded by some activity, situation, reading, or video that students respond to. During the fifth-grade unit on sound we mentioned earlier, the teacher showed a video of a singer breaking a glass with just the energy from his voice. She wrote on the board: "How can a person shatter a glass just by singing?" In the conversation, she then posed three kinds of questions, all of which apply to any eliciting experience:

- What did you notice happening here [before, during, or after the glass broke]?
- What did you think was going to happen in this [video, situation, demo]?
- What experiences have you had with [really loud sounds]?

When probing, the teacher is not evaluating student responses, but rather trying to get as much of everyone's experiences and initial ideas into the conversational space as possible.

Follow-ups

Follow-up prompts, as the name suggests, come after a student has made a comment and are requests for something additional. Follow-ups are just as important as the original question because they make students realize that you genuinely want to take time to hear more of their thinking. This means that you should be prepared to ask questions in pairs—an initial probe, then a follow-up. Consider how each of these follow-ups could be paired with one of the preceding probing questions:

- Can you tell me more about that?
- Can you explain/describe it in a different way?
- What do you mean by that?

The follow-up prompts work remarkably well for helping you understand "off-the-wall" comments by students. In a high school physics class, as we mentioned previously, one of our teachers was starting a unit on force and motion and shared a video of a young man doing urban gymnastics. He ran up to the side of a brick building, put one foot on the wall as he jumped, and did a backflip to land neatly on his feet. Before the teacher could ask, "What did you notice?" one student burst out: "He's got no fear!" For a moment, the teacher considered ignoring the comment and moving on, but instead she asked, "Can you tell me more?" The student hesitated, then explained that if the man had slowed down when he ran up to the wall, he would not have been able to make the flip. Other students then eagerly added a string of related ideas to this remark, using everyday language like "getting his speed up" and "getting a grip on the wall," which later in the unit were interwoven with the science language of momentum, unbalanced forces, and static friction.

In the classroom of a less experienced educator, this student's initial comment would likely have elicited a smile or been dismissed as irrelevant. In this class, however, his comment was followed up by the teacher, reworked by his peers, and later used as a resource for the whole class in explaining why the backflip was successful.

Pressing

There are times when a teacher must ask students to reason further (and out loud) about something they've just said. This is known as pressing. Pressing students is very different from eliciting their ideas and experiences (i.e., different from probing). When pressing, the teacher does not allow students to offer shortcut responses, unsupported claims, or the tagline "You know, right?" Instead, when a student offers an initial idea, the teacher might ask in return: "Why do you think that?" or "Isn't that a contradiction to what you said earlier?" or "What evidence do you have for that claim?"

Notice that these are not like generic follow-ups in which you want students to express more of what they notice or know; presses require students to go deeper to link, justify, or critique ideas. These are reasonable ways of holding students accountable for thinking. You can tell you are pressing because young learners will often visibly squirm in their seats when you won't give up on their thinking. Pressing can be done in whole-class episodes or as the teacher visits groups of students around the room. We note here that it is possible to press the whole class to reason further about a statement an individual has made. Next we'll identify some types of pressing moves by teachers.

Asking for examples

- Can you give an example of the idea you just mentioned?
- Can you think of a case where this holds true? Or does not hold true?

Pressing for consistency with other ideas or information

- Does your claim fit with the data we have?
- How is your explanation different from the one that [a peer] just offered?
- But is what you are saying consistent with [a known science principle, fact]?

Asking for evidence or justification

- What makes you think that?
- How did you arrive at that conclusion?
- What evidence do you have?
- How does that idea support your claim?

Requesting that students "fill out" an explanation

- Sounds like you have the start of an explanation [repeats students' partial claim or explanation], and you have the end, but isn't there something that happens in the middle?

Asking how one could test a claim or hypothesis
Occasionally a student will state a claim that you might recognize as being testable in a simple way. So, with the appropriate degree of support, you can ask the student, or the whole class:

- That's an interesting idea; is there a way we could test it to see if it's true?
- What resources might we need to do that?
- What would make a fair test?

The following example of testing a hypothesis is from one of our middle school lessons on cellular respiration. A group of students had just completed an activity in which they mixed yeast with some sugar and warm water in a flask. They then affixed a balloon on top of the flask and watched it inflate over a thirty-minute period. Some students explained to the teacher that the balloon was inflating because of warm air rising in the flask. Another group of students came up with a different idea. They thought that the yeast was "coming alive" and "eating" the sugar, then giving off carbon dioxide. The teacher, at the end of class, asked students: "Is there a way to test one or both of these hypotheses? Is there a way to test to see if one of the hypotheses is *not* true? What equipment would we need and what kind of data would we collect?"

Asking a "what if" question
You can get students to reason more deeply about a hypothesis or model by asking what would happen if part of a system they have been studying is changed. In the yeast and balloon vignette just described, the teacher might ask the first group who believed the inflation was caused by warmed air:

- I hear what you are saying, but what if we let this flask cool down for a few minutes—would the balloon still be inflated?

  More generic forms of this move are:

- According to your hypothesis, would we see something different if [condition X] was changed?
- Does this always happen this way; what might make the outcome different?

Using Wait Time

During whole-class discussion, students need time to think. Not everyone can spontaneously interpret what a teacher's question means and respond to it within a couple of seconds. Rapid-fire questioning advantages those few students who have mastered English, who can anticipate the types of questions the teacher will ask, and who can recall facts easily. Yes, they're the students who always have their hands up. The majority of your students, however, may be sidelined by the fast-moving dialogue, reduced to the role of spectators.

One way to make conversations more equitable is to pay attention to your *wait time*. This is the amount of time between when a teacher poses a question, and when the teacher either calls on a student, rephrases the question, gives a hint, or answers the question himself or herself. It is essentially the amount of time the teacher gives the students to think. Research has shown that the average wait time teachers give students is remarkably short—approximately one second. This is because teachers are almost immediately uncomfortable with silence in a classroom conversation, and they seek to fill the void with a student's voice or their own. Not surprisingly, short wait time by teachers is associated with the I-R-E pattern of discourse. This same research has shown that when wait time is kept short, only a small minority of students respond, and their responses are brief.

Wait time is something you can control. Extending these pauses is one of the simplest but most effective ways to encourage equitable and higher-quality participation in classroom discourse. Some teachers have purposely lengthened their wait time to five, ten, or twenty seconds, to give everyone time to think. In these classrooms, a *far greater percentage* of students responded to the teachers' questions, and the responses were *longer and more thoughtful*. We know it seems strange to use silence as a "talk move," but pauses in conversations can prepare everyone to contribute and be more thoughtful about their responses.

Opening Up Cross-Talk

One feature distinguishing Ambitious Science Teaching from traditional instruction is the use of peer-to-peer talk, specifically students' commentary on the ideas of their classmates. This requires particular discourse moves and norms for civil exchanges (described in the next chapter). Here are some types of questions that get students to compare, critique, and add on to the ideas of others:

- Can you rephrase what Marie said in your own words, and check with her to see if that's what she meant?

- Can you tell Jordan whether you agree or disagree with his statement and why or why not?
- Does anyone want to respond to that idea? And please talk to the person rather than to me.
- Can you two compare the ideas you just came up with? Are you saying they are the same thing?

As you can imagine, knowing who to ask these questions of, and when, requires that you build a trustful relationship with your students and that you establish a safe classroom environment for these conversations. It may be helpful to first try out these moves in small-group work; then, after students get acclimated to these requests, you can make them part of whole-class discourse. You will need extra patience with these moves in your classroom, since very few students experience talk like this in their other classes, or in their everyday lives. And remember, by themselves these moves do not get students to talk directly to one another, without you as a constant intermediary. Special norms and routines have to be put into place for this to happen (as we'll discuss in chapter 4).

Revoicing

Revoicing means that the teacher listens to an extended statement a student has made, then paraphrases and rebroadcasts to the class what was said. There are several reasons a teacher might do this, but regardless of the circumstances you must use this move with care because students might interpret it as an attempt to "overwrite" their ideas with your own. Revoicing is usually done in the context of whole-class discussion, but it can be used in visits to small groups as well. Following are some examples of revoicing and the reasons you would use them.

Revoicing to mark a segment of a student's idea as particularly important

Here's the situation—a student has taken a longish turn at talk to express an idea, a disagreement, or a personal experience that is relevant to the science topic. The teacher recognizes that one part of this lengthy response could be particularly helpful as a resource for the rest of the class to reason about. This is when the teacher could select that specific portion of the student's comments to restate. In discourse terms, this is often referred to as *marking* or *elevating* an idea. It may sometimes be only a word the student has said, but it can also be a full hypothesis, an observation, or a question.

Here is an example of a sentence stem a teacher might use after a student has given a legitimate, but long and occasionally disconnected, interpretation of a classroom demonstration: "So [name of student], what I hear you saying is that heat has something to do with the motion of the molecules of water in our food dye demonstration. Am I interpreting that correctly?" This discourse move is also used when there are a number of ideas flying around the room that could confuse students or that divert from the main idea that the class is working on. You could say, "We've just had a number of different contributions, so let me try to revoice what I heard."

Revoicing to repair how an idea is expressed

This revoicing technique is a restatement of a student contribution in which the teacher judiciously selects and interprets one ambiguous aspect of an otherwise valid claim or observation. A sentence starter here might be: "Here's how I understand your explanation [restate it], but did you mean to say ____, rather than ____?" Instead of "correcting" statements on a routine basis or evaluating them overtly, the teacher makes clarifications in a sensitive way to avoid confusing other students.

Revoicing to connect students' everyday language with academic language

Students often need to hear how some forms of everyday language they are using connect with scientific terms that sound similar. An example here might be: "So when I heard you talk about acceleration, you were using it to mean speeding up, like you do when you press on the gas pedal in a car. Scientists, though, use that term in a different way, to mean any change in speed—even slowing down—or a change in direction."

Another kind of connection between everyday and scientific (or academic) language is when a student uses a common term and the teacher revoices by substituting a scientific term—for example, adding "convection" to a student's description of "warm air rising while cold air sinks." The teacher here, however, must take care to maintain the student's ownership of that idea. We don't suggest that such statements, by themselves, help students fully understand the links between everyday and scientific language, but these are opportunities to help students see that words can have both disciplinary and common meanings—that is, that there are specialized terms for events or things in the natural world.

Focusing

Occasionally students will be overwhelmed by a question or task in a way that teachers could remedy by drawing their attention to a small part of the problem space. This can happen in whole-class conversation or in small groups. For example, if the students are working on an explanation, and there is a gap or an inconsistency in their account, you can say:

- Tell us just about [this smaller part] of the story—focus on that.
- I think you need to consider the [X] activity we did and think about what it told us might be happening here (pointing perhaps to a place in the students' written explanation or on their drawn model).

This move could also be used when students are interpreting a representation like a graph. In this case, rather than having them start by trying to make sense of the whole figure, you can ask:

- Tell me what you think these axes mean.
- If we take just one data point (or one bar on the bar graph) instead of all of it, what does this one observation represent? Use the labels on the axes to help you.

You can combine these focusing efforts with follow-ups like "Tell me more," or "Do you agree with what she or he just said?" There are many other situations in which simple requests to "Look at this part" or "Think about how [X] is related to [Y]" are helpful. Focusing can be powerful because you are still letting the students do the reasoning, but guiding them about where to concentrate their attention. We offer an important point of clarification here: Funneling is not focusing. Funneling is leading students to say a particular word or phrase (which is merely an I-R-E interaction in disguise). Focusing stimulates reasoning and allows for many kinds of responses.

Putting an Idea "On Hold"

In the enthusiasm of whole-class or small-group discussions, students often make statements that can be off-topic or would be better addressed later. In these cases, teachers need polite ways of acknowledging the student's contribution, while at the same time marking it as something that is not going to be explored now. A teacher might say: "That's an interesting idea, and it is something that we will talk about tomorrow, but for now . . ." or "I like your thinking, but let's hold on to that thought."

Some teachers have a section of their wall space devoted to genuine questions or comments that students have, which may not be the focus of the current lesson. This has been called the *parking lot*, and it signals to students that their ideas have value, but may not fit the current discussion. Do keep in mind that if you have a parking lot, then you *must* address these questions and observations at some point with students. Some teachers use the parking lot to avoid really interesting and relevant questions from students. Do not use this strategy to suppress students who want to contribute what they think are important ideas.

Some Assembly Required: Using the Discourse Moves Toolkit to Be More Responsive to Students' Ideas

Looking across all the types of talk moves outlined in figure 3.2, you can see that there is no apparent order or frequency with which they might be used. In a whole-class conversation, for example, the only move you would be sure to use is the initial question; the other moves have to be selected in response to what students say or don't say. It might sound daunting to be so improvisational. But if you can identify a goal for the conversation, you'll be better able to determine the range of science ideas that are relevant to the conversation, imagine responses your students might give, and subsequently narrow down the possible moves you'd use. Are you trying to elicit students' initial ideas about a phenomenon? Then you might use a sequence like this: probe > follow-up > probe > follow-up > wait time > revoicing. Are you asking students to improve their current explanations with evidence from a recent activity? The sequence might begin this way: press > wait time > focusing > follow-up > press > follow-up > opening up cross-talk. These represent just the first few minutes of a conversation, and just one possible arrangement of talk moves. If you look back to the example conversations earlier in this chapter on eliciting students' ideas about solar eclipses, you can see the pattern of talk moves Teacher C makes that allows her to understand a great deal of what her students currently know and how they reason about the topic.

Experience will also help you become more adept at selecting talk moves. When you first do this work, you'll feel like a novice soccer player in the middle of a game, trying to pay attention to everything at once, unable to predict where the action is likely to go next, and trying to decide among a dozen different moves when it's your turn to handle the ball. With time, however, you can

FIGURE 3.2 **Talk moves and ways to use combinations of them to accomplish goals for conversation**

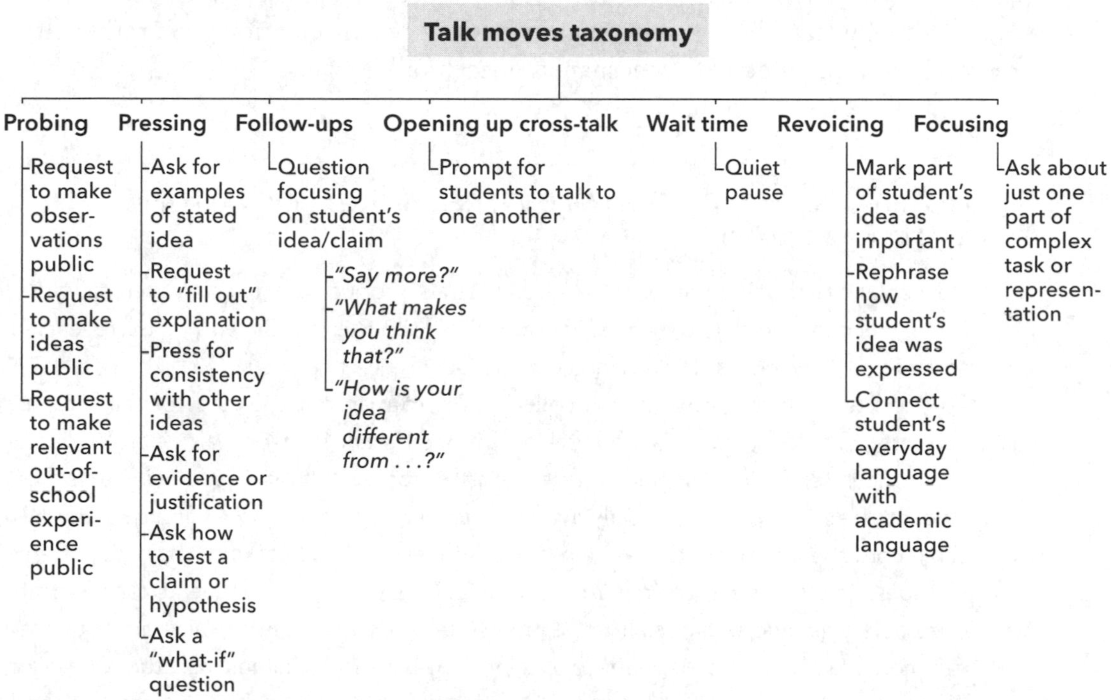

Talk moves taxonomy

Probing
- Request to make observations public
- Request to make ideas public
- Request to make relevant out-of-school experience public

Pressing
- Ask for examples of stated idea
- Request to "fill out" explanation
- Press for consistency with other ideas
- Ask for evidence or justification
- Ask how to test a claim or hypothesis
- Ask a "what-if" question

Follow-ups
- Question focusing on student's idea/claim
  - *"Say more?"*
  - *"What makes you think that?"*
  - *"How is your idea different from . . .?"*

Opening up cross-talk
- Prompt for students to talk to one another

Wait time
- Quiet pause

Revoicing
- Mark part of student's idea as important
- Rephrase how student's idea was expressed
- Connect student's everyday language with academic language

Focusing
- Ask about just one part of complex task or representation

If goal is eliciting ideas and activating prior knowledge, use combinations of:

- **Probing** (e.g., "What experiences have you had with . . . ?")
- **Gentle follow-ups** (e.g., "Can you tell me more?")
- **Wait time or time to write privately**
- **Revoicing** (e.g., "So what I hear you saying is . . .")
- **Opening up cross-talk** (e.g., "Does anyone want to respond?")

If goal is small-group sense making about observations or activity, use combinations of:

- **Probing** (e.g., "What did you notice happening here?")
- **Challenging follow-ups** (e.g., "What do you mean by that?")
- **Revoicing** (e.g., "Here's how I understand your claim . . .")
- **Focusing** (e.g., "Let's just talk about this part of the model . . .")

If goal is for whole class to connect activity with "big" science ideas, or press for explanations, use:

- **Pressing** (e.g., "Does your explanation fit with the data?")
- **Challenging follow-ups** (e.g., "How is that different from what was just said?")
- **Wait time**
- **Opening up cross-talk** (e.g., "Who agrees with this and why?")

better anticipate where a classroom conversation is going, focus your attention on just a few aspects of the talk that really matter, and more easily recognize when you have to use particular moves to accomplish a goal. In short, you'll eventually learn how to focus on the flow of the conversation itself rather than on each moment and the mechanics of each talk move.

HOW TO GET STARTED WITH DIFFERENT KINDS OF TALK IN YOUR CLASSROOM

How does one develop the expertise to orchestrate productive talk on an everyday basis? For any teacher, there are principles of responsive discourse that can be studied, rehearsed, and refined. Expertise is simply a matter of deliberate practice.

One place to begin work with colleagues is to map out the kinds of talk that are currently going on in your classrooms. How many different students participate in whole-class discussions? How long are your students' responses? If you get blank stares from students frequently, this is actually a great opportunity to use data to find out why. You can record the questions that result in stony silence and the context in which you offered the question. Trends will emerge. You might find, for example, that students can't hypothesize about generic situations (like weather systems described in the abstract), but they can more readily start talking about a specific contextualized example (like the windstorm that occurred last week and the damage it did in the neighborhood). Events that are local, personal, or part of everyday experience are always more accessible to students as starting places for more demanding questions.

Your data collecting might reveal that you've asked a question to start the class, but without orienting students to how it builds on an activity done the previous day. Or you may find that students do not have the background experiences to understand what the question is asking. One of our teachers asked students, "What are the types of energy required to bring a roller coaster from its starting position to the top of a track?" only to realize later that many of his students had never been to an amusement park.

Once you have baseline data on students' contributions, turn to examining the choices you are making during class. Ask colleagues who are inquiring with you, for example, "Do I make the goals for discussion explicit for my students? If so, how?" Record what is actually said when you attempt to frame a discussion. Try to see if your discourse moves work to support the goals of the conversation. As you test out new discourse strategies, you should always track data on how many students participate and the types of their contributions.

FOUR

PRODUCTIVE DISCOURSE, PART 2

Encouraging More Students to Participate in Talk

IN THE PREVIOUS chapter we described discourse from the teacher's point of view—identifying goals for discussion, selecting challenging questions to start a conversation, and using talk moves to support student reasoning. In this chapter we consider the students' experience, specifically how you can help students feel capable of participating in science discussions and willingly share ideas with others. This requires "laying the groundwork" for discourse by establishing conversational *norms* with students that create an atmosphere of civility and personal safety, by developing everyday *routines* that they will recognize as opportunities for particular types of talk, and by designing *scaffolds* that help them find the words to sustain discussions in different settings.

Together, strong norms, predictable routines, and strategic scaffolding enable talk to flourish and make each student less likely to think, "I can't participate" or "I won't participate." There is a difference between these two responses. Students often convince themselves they *can't* participate in discussions because they lack the prerequisite science knowledge or they don't know the rules of the game for a particular kind of talk (for example, debates about the use of evidence or what the teacher means by "unpacking" an idea). Some students cannot process language fast enough to interpret what peers are saying, then find an entry point in the conversation for their own ideas. On the other hand, students may feel capable of contributing to a discussion but *won't* because they fear their ideas will be unfairly criticized, or worse, not taken seriously. Occasionally, students can also feel they are not being intellectually challenged and become too bored to participate. You can problem-solve most of these issues by first considering whether students' reluctance is a case of "can't" or "won't," and then adjusting the classroom norms, routines, and available scaffolds accordingly. As you'll see in this chapter, there are lots of options, and they work best in combination with one another.

NORMS TO ESTABLISH AN ATMOSPHERE OF CIVILITY AND SAFETY

Productive conversations require students to take risks in public—to hypothesize about science ideas they are only partially familiar with, to comment on the contributions of classmates, or to ask questions that may reveal a lack of understanding. Because of this, the most basic prerequisite for productive conversations is that all students *feel safe* in speaking out. A safe classroom is one in which students will not have their ideas ridiculed, and their teacher and peers will value what they have to say. These conditions require explicitly stated norms for civil interactions.

It is helpful to co-construct these norms with students, but we recommend that you select/construct at least one from each of three categories: respect for others, equitable participation, and accountability to the science as well as one's classmates. If you can decide on no more than four or five norms in total, students will be more likely to remember and use them. Table 4.1 shows samples of norms that our teachers have developed with their students (adapted from Suzanne Chapin et al.).[1] Some of the phrasing is aimed at high school students, some for middle school or elementary students, but any of these norms can be translated for younger or older audiences. There are many possibilities, but keep in mind that this list includes *more* items than you would want in your final set.

Set your norms early in the school year. In the first days of September, the teachers we have observed negotiate with students how they will treat each other through talk. You'll want to demonstrate, with the help of your students, what each of the norms in your list sounds like or looks like in conversations. You can role-play and put students in charge of coming up with positive or negative case examples of each. In addition to helping students understand them, it is essential to *enforce* the norms when they are at risk during student discussions. For example, you might have to step in and say, "That's a put-down

TABLE 4.1 **Possible classroom talk norms by category**

MENU OF EXAMPLES TO SELECT FROM	CATEGORIES
Preparation: We come prepared for discussion with notes, examples, stories, and our readings.	Accountability to science and classmates
Responsible learners: We are responsible for our own learning. This means we speak, request clarification, show agreement or confusion, verify, and ask others to repeat.	
Pushing ourselves: We push ourselves and each other to think beyond the obvious, disagree with ideas, and draw out comments from classmates, and we are open to changing our minds.	
Focus: Our comments and stories will stay on topic, and we have the right to explain how our contribution connects with the science.	
Hearing from all: Everyone deserves to be heard.	Equity
Air time: Don't monopolize the conversation, a.k.a. "Watch your air time."	
Priority to newcomers: We'll give priority to those who have not had chances to talk yet.	
Time to think: The teacher will give "think time" before asking for our ideas.	
Civil participation: No put-downs—ever.	Respect for each other
Impulse control: Don't interrupt or talk over your classmates when they have the floor.	
Fair critique: We, students and teacher, can critique ideas of others, but personal attacks are out of bounds.	

rather than a fair critique of an idea; how can you rephrase what you said?" You can also call out positive examples of adhering to particular norms: "Thanks for inviting others to get into the conversation."

It can be helpful to ask your students what might keep them from participating in a discussion. Explore with them the ways that people can feel slighted or attacked when they express an idea. What are some comments that classmates might be bothered by? How can we avoid those? And what should we do if someone makes a critical remark or says something disrespectful? A scholar of classroom discourse, Cathy O'Connor, suggests a direct conversation with students:

> Because so much of the school experience is focused on producing right answers in the form of a one- or two-word response, your students may be puzzled or uncomfortable in a talk environment where they are asked, for example, "Why do you think that?" . . . Some students might believe that you are telling them they are wrong and need to keep talking until the correct response comes along. This gives you a chance to explain why probing and pressing kinds of talk represent a useful and positive exploration of ideas—not a sign that they are wrong.[2]

In addition to up-front discussions about the value of norms, you can periodically gather data on how students feel their ideas are being received by others. After class conversations, you can ask: "How did we do today in our discussion? What do we need to work on?" Students, however, won't jump at the chance to answer these questions, so you can use *exit tickets* to get input from everyone. Exit tickets allow students to give feedback—anonymously, if they prefer—on specific aspects of their classroom experience. The example exit ticket shown in figure 4.1 gathers data from students by asking them to evaluate their small-group work on a scale from 1 to 5. Sometimes you'll get

FIGURE 4.1 **Student exit ticket to assess participation in talk and "being heard"**

Student exit ticket

Name (can be optional) ____________________ Date __________

On a scale of 1 to 5, please circle how you felt in science class today:	**Not at all**		**Some-what**		**A lot**
I felt my ideas were listened to or heard by my group.	1	2	3	4	5
I felt my ideas were understood by my group members.	1	2	3	4	5
I felt other students were interested in hearing my ideas.	1	2	3	4	5
I felt that sharing my ideas helped others think more.	1	2	3	4	5

encouraging feedback, while other times the responses might help you recognize that particular groups of students are feeling marginalized.

Norms are "alive" in classrooms and need modeling, enforcement, and check-ins with students. Early in the school year, talk time will be marked by uncomfortable silences and by you being really explicit with students about "what counts" as productive interactions around ideas. Much of the dialogue that will appear so natural later in the year will have been established by many weeks of students' clumsy attempts to interact with each other, and by your steady reinforcements of respect, equity, and accountability.

ROUTINES THAT ENCOURAGE LISTENING AND PURPOSEFUL TALK

Establishing routines is another fundamental part of laying the groundwork for productive student discourse. Students understand routines—they quickly pick up on teacher signals like making a familiar type of announcement, asking for a rearrangement of the classroom space, or sometimes just gesturing in a certain way. Predictable routines help students understand what their roles are in an activity and prepare them to participate. Teachers can take advantage of nearly any part of a class period to introduce well-structured routines that involve talk. We have seen colleagues transform mundane procedural tasks into rich opportunities for student discourse—for example, during warm-ups in the first five minutes of a lesson, as part of small-group work, and at the point where ideas are summed up at the end of class.

We provide several examples here, but these are not the only routines possible; in fact, there are dozens of variations. Our purpose is simply to demonstrate what routines are and how they can facilitate learning through talk. You'll notice that scaffolding is frequently built into these routines, and that the success of routines depends upon a shared set of talk norms. It's a reminder that norms, routines, and scaffolding reinforce one another.

Having Students Think-Pair-Share

In this common routine the teacher poses a question and asks students to *think* silently for a minute or two how to respond, then *pair* with a peer to compare ideas, and finally return to the whole-class conversation to *share* their thinking. The following example was used during the five-minute warm-up of an introductory lesson in eighth grade relating to the gas laws:

> Okay, let's get our notebooks out. Here's our starter question [this signals a routine]. At the end of yesterday's class, we all agreed that air is made up different types of gases. Oxygen is one example we're familiar with. What I'd like you to do is take a few moments, by yourself, and write down another example of a gas that you think makes up the air around us. If you can think of more than one, that's great.
>
> [After 60 seconds] Now I'd like you to pair up with a partner and share what you have. Then together, pick one of your examples and talk about what convinces you that this gas is really part of our air.

In this example, the teacher poses an initial question that activates students' prior knowledge, asks them to think a bit more deeply about their common understandings of "air," and allows for lots of possible responses that can be drawn from everyday experiences. Imagine if the teacher had instead asked what the definition of a gas is. Rather than probing their own background knowledge, students would be fumbling through their textbooks trying to copy a few lines for the teacher's approval. When you ask accessible questions, however, students will likely generate a diverse range of ideas to share when they pair up. Then, they get to hear each other's thinking and get practice at expressing themselves without competing against twenty-five other voices.

In our example, the teacher has added a further challenge for students—to declare *why* they think a particular gas makes up part of our air. When everyone is brought back together for whole-class sharing, students can try out stating a claim ("Carbon dioxide is in the air") and offering some evidence ("We breathe it out" or "Cars create it when they burn gas"). Other students can be asked to comment or ask questions in response. You can leave out the academic labels of "claims" and "evidence" in the initial phases of such discussions, then introduce them later as useful terms that apply to moves that students have already made. This routine, then, as brief as it is, can be used to acclimate students to scientific argumentation. Arguments do not always have to include formalized rhetoric with graphs and charts to support one's claims. Rather, they can simply entail saying what convinces you that some assertion about the natural world is true. Students' everyday knowledge and experiences count as resources for these mini-arguments. During this warm-up routine, every student in the room is asked to think and has a chance to "talk science," even if briefly.

Using Sentence Frames to Generate and Revise a List of Hypotheses

In nearly any unit you teach, there can be an opportunity to generate, with students, a list of hypotheses for some science phenomenon (for example, alcohol evaporating faster than water, the formation of solar systems, or animal migration). You can start these activities by saying to students: "We are trying to lay out the possibilities for what might have happened here and how it happened. Many different ideas are possible, and we'll welcome everyone's hypotheses." You can offer some help to those who are shy about participating by writing *sentence frames* on the board. A sentence frame is simply a portion of a sentence that signals to students the basic format of a productive comment (written or spoken). The student then fills this in with any number of possible remarks. It is not intended to funnel students into saying a specific word or phrase.

One example might be: "We think ____ might have something to do with it." Another is: "We think this event is caused by ____." The former signals to students that a condition or situation might be all they are required to comment about, with no expectations to describe a causal story. Participation, then, seems less risky. The latter example requires an explanation, a bit more speculation and risk. Some students will prefer to use the first of these frames, others will prefer the second, and still others will create a response without the frames; it doesn't matter as long as they know what you are asking for. The resulting list from students is usually a mix of one-sentence observations, inferences, and mini-theories. They are not full-blown explanations.

As the list is built, students (usually working in pairs) know they can either offer new hypotheses or comment on one that is already on the list. You can probe a bit when a statement is offered, and occasionally you might place a question mark behind a hypothesis so students understand that it is not yet supported by evidence and may even be inconsistent with science. Students become aware later in the unit, as they engage in cycles of reading and activity, that they can gather evidence and ideas that can be used to test the hypotheses in their list. Some of these hypotheses may get crossed out as implausible, some might be supported or elaborated upon as time goes on, and some might be combined with others to make more sense. Because this routine requires that students comment on the ideas of their peers, you'll want to reinforce norms of respect and civility in discussions.

"Talking About Talk" After Students Share Work

Several of the teachers we work with use a routine in which a student is selected to display a scientific model (drawing of an event or process) he or she has been working on. This student is identified during a think-pair-share exercise in which partners are asked to talk about changes they made to their model the previous day. As our teachers circulate during the pair conversations, they listen and watch for changes in students' models that might stimulate conversation for the rest of the class. Maybe the changes are not consistent with science, or maybe the models are simply drawing attention to some aspect of the focal phenomenon that the rest of the class had not yet noticed.

In one recent case with fifth graders, the previous lesson focused on how bacteria appeared to multiply on a slice of potato over a period of days, despite an early application of an antibacterial spray. The student that was selected had theorized that the mixture had killed off only some bacteria, and that there might be a whole "zoo" of organisms that could grow on the potato, not just those that were susceptible to the spray.

In this routine, the presenting student places the model on the document camera so everyone can see it. The student can, and usually does, ask her partner to come along. The student's role here is to explain what thinking and experiences led her to make changes in her model. She lays out her rationale, then has the right and obligation to call on peers who want to ask about the model. The rest of the class has a well-understood repertoire of responses they can offer, including posing clarifying questions, asking what evidence she used to make changes, and stating how they are viewing the presenter's ideas in comparison with their own. After the routine is finished, our teacher has a meta-talk with students—in other words, a talk about talk. He asks, "What did we do well today?" Students reference a nearby poster that lists moves like "Asking genuine questions," "Using representations to make a point," "Requesting more information," and others. His students direct him to place a sticky note on each kind of talk move that they or the presenter used.

In this routine, students are authorized to present and question ideas, then to assess their own talk in order to continue improving what and how they learn together. It's worth noting that in the student-led discussion about the bacterial "zoo," the class engaged in dialogue for over five minutes without the teacher saying a word. This is actually a long time for students to sustain a conversation without the teacher—you may want to try this routine and see if your results are similar.

Providing Equitable Access to Information and Ideas During These Routines

For any routine in which you expect students to talk, you have to consider whether they need shared knowledge or information to participate. This is why whole-class conversations, for example, can falter—some students might have access to relevant ideas and experiences to use as resources, while others do not. Most equitable types of talk routines require that all students have read, listened to, viewed, or manipulated something as a prerequisite for engaging in the subsequent conversation. This, again, is why there is no such thing as a discussion in which the teacher simply says, "Let's rap about chemical equilibrium." Unless everyone has access to some basic knowledge and experiences, there can be no level playing field for the talk.

SCAFFOLDS FOR TALK

Scaffolding, our third component for fostering productive student discourse, means providing temporary forms of assistance for learners who are faced with instructional tasks that are just beyond their current capabilities. Basic scaffolding moves can include:

- *structuring* the task itself (teacher provides some order to follow or way to organize the work to be done) without making it overtly procedural;
- *focusing* learners temporarily on particular parts of a larger task rather than on the complex whole;
- providing specialized *tools*, like sentence starters or sample questions one can ask peers during group work;
- teacher *modeling* or "thinking out loud" about how to approach different aspects of a task;
- *coaching* learners as they work on tasks; and/or
- providing timely *feedback* about learners' thinking or participation.

In scaffolding, the *goals of the task itself are not simplified*; rather, the roles learners are asked to play and the supports they are given in completing the task are modified, depending upon their current levels of understanding and skill. We already showed one type of scaffolding in our previous example of constructing a list of hypotheses: students were given sentence frames to help jump-start their verbal contributions. The overall task remained demanding, but these specialized tools provided assistance.

The big picture of scaffolding is like an apprenticeship. In the real world, when people are learning to become pilots, physicians, plumbers, or architects, they get different kinds of assistance as they try out selected parts of authentic tasks. Novice pilots fly in a simulator and then act as copilots, taking on peripheral jobs in the cockpit while the more experienced captain coordinates decisions required to keep the plane flying. Medical interns, journeyman plumbers, and architects-in-training all take on structured but limited tasks, get coached on how to approach unfamiliar work, and receive feedback on their performances. For these individuals and for students, apprenticeships mean taking fuller responsibility over time as they become more independent problem solvers. The roles they play in complex tasks and the extent of scaffolding change, then, as learners become more competent. Scaffolding is "faded" or gradually withdrawn over the school year.

The Structured Talk Scaffold

Talking with another person about science ideas is actually quite complex in terms of the different conversational moves that one has to use and how responsive one has to be to the other person's thinking. Structured talk is a scaffolded version of the think-pair-share routine. We say scaffolded because it *structures* the turn taking and the roles that student use in the pair interaction. It allows students, working with partners, to each get chances to reason out loud, learn to listen actively, and revoice another person's reasoning. These are fundamental skills that do not develop naturally for most students.

You can use this strategy when you've posed a science question that requires students to talk about evidence in their response or to synthesize information to come up with a hypothesis or conjecture. You don't want to have your students think there are "right answers"; you want them to learn that there are multiple ways to reason about science ideas productively.

Start by selecting student pairs. One student is Partner A and one is Partner B. You may want to be strategic about partnering beginning English language learners with more advanced ELLs, and using other pairings in which one student may be a bit more capable than the other in terms of reasoning or language use. Then choose one of the following three options.

Option 1: Listen and revoice

1. Use private think time to sketch out some ideas.
2. Form A/B partners.
3. Partner A explains his/her ideas by using reasoning, representations, claims, explanation, or evidence. Partner B silently listens to understand A's thinking.
4. Partner B then revoices what Partner A's ideas were, without judging, adapting, or commenting on correctness of ideas.
5. Partner A clarifies what was said as needed.
6. Partner A and Partner B switch roles.

Option 2: Build on or challenge an idea (variation of Option 1)

1. Form A/B partners.
2. Partner A explains his/her ideas by using reasoning, representations, claims, explanation, or evidence. Partner B silently listens to understand A's thinking.
3. Partner B builds on or challenges Partner A's ideas using representations, explanation, and evidence.
4. Partner A responds to Partner B's ideas or challenges using representations or evidence.
5. This continues until consensus is reached or until both agree to disagree for the moment.
6. Partner A and Partner B switch roles.

Option 3: Identify similarities and differences

Follow the preceding protocols. Then, after Partner A and Partner B both have opportunities to share and interact with each other's ideas, they jointly discuss similarities and differences between their ideas.

Figure 4.2 is an additional scaffold for students using Options 2 or 3. It shows various sentence frames that a listening/responding partner could use. Our teachers often place a card like this on the students' lab tables so they are more likely to try them out.

FIGURE 4.2 **Sentence frames on a card to support structured talk**

Here's what I think. . . .

After I listen, what can I say?

I'm not sure I understood _____. Could you tell me more?

Can you repeat the part about _____?

What do you mean when you say _____?

I heard you say _____. What makes you think that?

I heard you say _____. What if _____?

Would you explain a bit more about _____?

Talk Role Scaffolds for Small-Group Work

How often have you put clusters of students together to discuss a reading, plan an investigation, or carry out a lab, only to later find them toiling away as individuals, having apparently taken vows of silence for the duration of the task? If we want students to interact with one another during small-group activity (three or four individuals), we have to design their opportunities to talk as intentionally as we design the activity itself. To foster intragroup participation and the talk that comes with it, it is helpful to assign roles to students. Roles should *not* be simply about managerial duties, such as the note taker, the supply getter, the procedure reader, or—our favorite—the trash captain. Instead, roles should direct students to take responsibility for different parts of science talk that moves everyone's thinking forward. We are not suggesting that students' only opportunities for talk are through these roles, but rather that they learn to "activate" the role as needed—this means that they'll need support for what to say and when to say it. Next we provide some examples of roles for intellectual work. We *list more here than you would assign* to a group of students, and you might combine these roles in some cases. What we show here is not "the list," but rather suggestions based on our experience in classrooms.

Big ideas person

- Often a group will get too wrapped up in the rote execution of directions for an activity. The big ideas person pauses the group and pulls them back to the scientific purpose of the activity (which should be made clear by the teacher). The term "big idea" is used here as a placeholder for goals like understanding a concept or interpreting data in order to support an explanation. A person in this role asks the following: What is the big science idea we are trying to understand?

- How does the work we're doing (something we are studying, reading, investigating, observing, etc.) help us understand the big idea?
- How does our work so far change the way we're thinking about the big idea?

Clarifier

This is a role for monitoring everyone's comprehension about one or two key science terms. This person might ask:

- Do we know what the words [like *carrying capacity* or *convection*] refer to in our activity?
- Can we put the term into our own words?

Questioner

This person asks probing questions during the activity. Students in this role also listen for questions posed by other group members and then revoice the questions to make sure that the whole group takes a moment to hear it. This is not a role that students find easy, so it helps to provide them with question stems:

- [Paraphrases what others have said]: So, what I think you are saying is _____. Is that right?
- [Asks in response to a claim by a classmate]: How do we know that?
- What would happen if we changed _____?

Skeptic

Often the group is working to produce something like an explanation, a design for an investigation, an argument, or a model. The skeptic's role is to strengthen the group's work by probing for weaknesses in the product being developed. This person might ask:

- Here's an alternative explanation—is this just as good as the one we have now?
- Why do we think that our [claim, model, explanation, argument, design for investigation] is consistent with the available evidence?

Progress monitor

This person asks others to periodically take the measure of the group's progress toward a goal.

- What can we say we've accomplished so far?
- What do we still need to know/do to accomplish this task?
- What can we now add to our [explanation, model, argument] that we didn't have before?

When you stop by a table to listen in on a group, you should expect this person to be able to communicate the ideas of the group members and attribute ideas to particular people (giving credit where it is due).

Floor manager

In group work, no one should be left on the margins of the conversation. Sometimes a student or two will dominate the talk and this is what keeps others from contributing. The floor manager monitors the airtime of group members. At specified times, they are allowed to manage who has "the floor," with the goal of ensuring that everyone gets a chance to talk and that everyone pauses to listen. They ask:

- Can we take a minute and hear from everybody before we move forward?
- Who has not had a chance to weigh in on this?

All of the aforementioned roles and their prompts reflect how highly functional groups actually operate in the real world. For example, people who work together productively in business, high tech, or the arts constantly orient themselves toward one another's thinking, clarify ambiguous ideas or plans, consider alternatives, and assess their progress toward explicit goals. In a classroom, playing out roles can be awkward at first, but these scaffolds support equitable opportunities to learn. After the initial hesitation of turning to peers and asking these prefabricated questions, students find that the activity gradually become more natural. They begin to spontaneously adopt the roles and improvise on the questions, in which case the scaffolds may no longer be needed.

In classrooms, these role cards are typed out and taped to desks or tables. Each card has a title, a brief description of the role, sentence starters for students, and even cues about when to enact these roles during the activity. On days when roles are first introduced in classrooms, we've seen our colleagues at the beginning of an activity call out one of the roles to be enacted in that moment. They do this again with a different role in the middle of the activity, then again with another role near the end of an activity. Students feel less sheepish about initiating questions of their peers when the teacher prompts them to do it.

Scaffolds for Whole-Class Discussion

In whole-class discussions, peer-to-peer talk happens when students begin addressing one another rather than the teacher. Peer-to-peer talk supports knowledge building and an understanding of how to argue scientifically; however, as with other types of classroom discourse, it does not happen naturally. Some students don't participate in whole-class, peer-to-peer talk because they are so unfamiliar with critiquing the ideas of others or adding onto one another's comments. They don't know how to publicly disagree with one another, and some students feel uncomfortable or that it is pointless agreeing with others in class.

You need two things to get peer-to-peer talk going in your classroom. One we have already described is establishing norms for a safe environment to offer and critique ideas. The other is scaffolding for the interactions students should have with each other's ideas. Together, these two conditions form the ground rules for public talk. Our teachers have developed, with students, lists of sentence starters that help them engage in peer-to-peer talk. We've discovered that four categories seem to work well for students: adding on to a peer's comment, asking for evidence or reasoning behind a claim, asking for clarification, and respectfully disagreeing with an idea. We share examples from a high school classroom in figure 4.3, which is one of four posters developed by students during the first week of the school year. The title, "Ask a Probing Question," is a variation on asking for reasoning and evidence behind a claim.

Will these tools immediately transform your whole-class conversations? No. Will you notice a difference in talk if they are used regularly? Yes. We've found that asking students to rehearse these kinds of exchanges gives them permission to use the sentence frames in addressing one another. It helps if you first use these frames during a whole-class conversation that follows an activity or reading, so that everyone has some shared experience to discuss. You'll also need to pose an opening question that will allow ideas to be voiced, compared, and elaborated on by students.

FIGURE 4.3 **Student-created scaffolds for addressing peers' ideas**

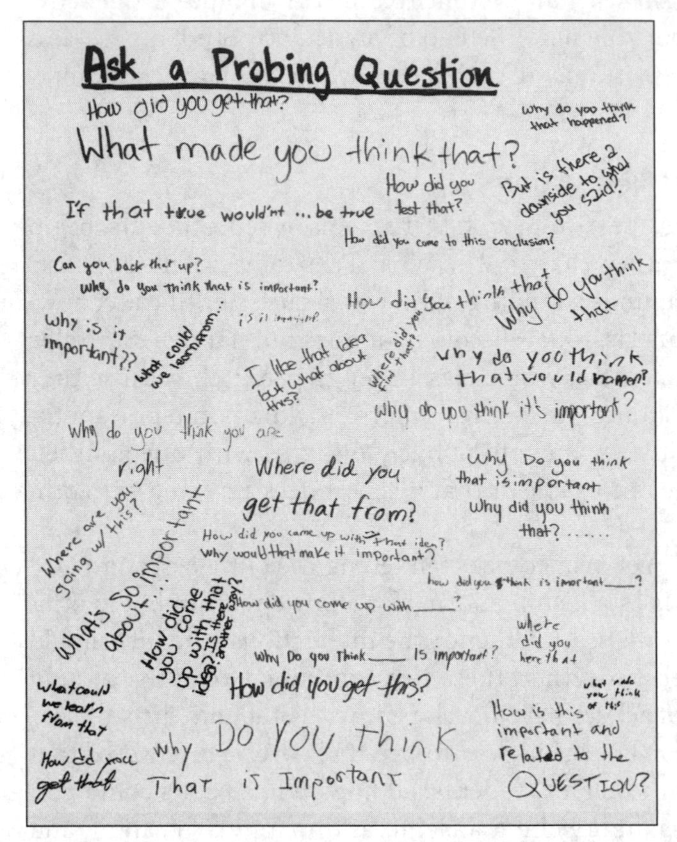

Example prompts from the other three posters in this classroom set:

Adding to an idea:

- I agree with you, but I also think . . .
- I agree with you, but couldn't you add . . .?
- I agree with you because . . .

Respectfully disagreeing with an idea:

- I know where you're coming from, but I have a different idea.
- I disagree with the idea because . . .
- I think you're heading in the right direction, but . . .

Asking a clarifying idea:

- What do you mean by . . . ?
- What makes you think that?
- Could you be more specific?

To further encourage peer-to-peer exchanges, some of our teachers move themselves to a corner of the room, signaling to their students that they need to carry the conversation. This gets easier over the course of the year with opportunities for students to practice and to occasionally, with your help, debrief the conversation.

Cultural Differences in Students' Expectations to Share Ideas

You will find that some students, despite your best efforts to encourage talk, remain steadfastly silent. In some cases, these individuals may hold culturally based expectations for how teachers and learners should interact around ideas.

In some (but not all) cases, students have had previous schooling experiences that have led them to believe that offering an idea of one's own to the teacher is presumptuous and inappropriate. They believe that the teacher dispenses authoritative knowledge, and their job as students is to listen and occasionally submit to questioning. These children can appear resistant when prompted for their opinions or asked to directly address a peer in whole-class conversation. We are not stereotyping here, but trying to raise awareness about cases that we have to be sensitive to. Similarly, some girls may appear to be "naturally shy" about contributing to science talk, but there is nothing natural about their reluctance. Many girls receive messages at an early age that adults and boys have the privileged voices, and that they should defer to them in discussions of any type. This trend cuts across ethnic, racial, and social class lines and can be reinforced by superficial images of young women in the media. We should not be surprised, then, when only a fraction of students in our classrooms feels comfortable joining in talk.

What can we do about these specific inequities in participation? On one hand, many students will find their voices if you use the strategies outlined in this chapter: establishing norms for a sense of safety, creating routines in which students understand the rules of the game for discourse, and employing scaffolds that give them "just enough" help to get into a conversation. On the other hand, we will also be dealing with students' fears, anxieties, and feelings of inappropriateness, even when inviting them warmly to say a few words. These are serious tensions for teachers who want to work with students' ideas, and our advice is to use small steps. For example, try to create opportunities for these students to talk in small groups, first by giving them meaningful roles to play in conversations. Place them with friends and allow them to try out their words with them (critiquing an explanation or asking a peer to clarify what they've said). The structured talk routines we discussed earlier are helpful in these cases. In whole-class settings, have them report out *with* groups, perhaps starting with them offering observations or consensus ideas; then, as the weeks go by, positioning them to share their own claims or questions; and finally, asking them to comment on what others have said.

We want to make clear that these individuals do not have deficits; they are as capable of intellectual work as anyone else in the room, and are simply dealing with emotional responses to public speaking at the same time they are learning to talk science. Some readers may argue that these recommendations will traumatize young learners, and we acknowledge that there is plenty of discomfort to go around when you try to make these changes. But we are thinking of the end game for these students—can we allow them to leave our classrooms without knowing how to communicate, to take a stand, to hear how others might agree or disagree with them? These are not just "nice to have" skills for science students—they are tools for participating in a democracy and for asserting one's self in society. Not helping students overcome their own silence would only reproduce inequities that are already too commonplace.

HOW TO GET STARTED

When you coordinate the use of norms, routines, scaffolds, and discourse moves, you increase the probability that a wider range of students will be able to participate in whole-class discussions and small-group talk. You also increase the likelihood that the talk will be productive. Success, however, requires *deliberate experimentation*—figuring out what aspects of discourse you want to work on with students, and then being methodical about trying out particular kinds of talk with them or using specific strategies for supporting that talk. Then you can assess, in a more targeted way, how your students respond.

Consider the checklist in box 4.1. It lists conditions that lead to a discourse environment. Select one of the items you could focus on for a few class periods. With some of your colleagues, decide on some observable metrics for assessing the impact of explicit norms, new routines for talk, scaffolds, or discourse moves by the teacher or by students themselves. The exit ticket we showed earlier in figure 4.1 is a good way to collect data on who is participating, and perhaps as importantly, which groups of students feel that their ideas are not being heard.

Developing peer-to-peer talk is particularly difficult work that will take weeks or months of encouragement. Most of our teachers who are experienced in classroom discourse look for students to begin engaging with one another after about three months. Yes, that's a long time, so persistence again is key.

BOX 4.1 Checklist for a safe and productive talk environment

- ☐ Norms for talk are co-constructed by students and teacher and reinforced regularly.
- ☐ Goals of classroom discussions are anticipated by the teacher and made clear to students.
- ☐ The teacher signals the start of recognizable talk routines that allow students to take up clear roles around the sharing and critique of ideas.
- ☐ Students' puzzlements and ideas are treated as resources for the learning of the whole class.
- ☐ A variety of discourse moves are used to manage the elicitation and development of ideas from all groups of students.
- ☐ Strategies allow students time to think during whole-class and small-group discussions (wait time, private writing time, think-pair-share).
- ☐ Students' everyday ways of expressing science knowledge are recognized as legitimate, but the teacher also makes links between these expressions and more scientific language.

FIVE

CORE PRACTICE SET #2

Eliciting Students' Ideas

WHEN USING THE AST FRAMEWORK, you begin every series of lessons by finding out what your students already know about the science you will teach. Two powerful principles from the research on learning tell us why we should launch units this way. First, people of all ages exert great effort toward sense making about novel stimuli in the world, and they will use anything at their disposal to do this work. In classrooms as well as in everyday life, students compare what they already know with unfamiliar ideas, and test whether new ideas being presented by the teacher somehow fit, or not, with their current mental schemas. Second, even the youngest of learners possesses a wide range of existing conceptions, prior experiences, and information that they use to make sense of any and all ideas introduced in the classroom. This prior knowledge takes the form of partial ideas that accrue from previous instruction, conversations with friends and family, everyday observations around the neighborhood, and stories in the media. We refer to all of these as *resources* because they have the potential to support learning. Students will, in fact, rely heavily on them in the classroom, whether you are aware of it or not.

Some of students' resources are useful (like outside-of-school experiences with evaporation or knowledge of inherited traits in their families) and a few can be problematic (like the belief that trees get their mass from the soil they grow in), but research clearly indicates that if teachers do not find out what these resources are, students will continue to use them to make meaning of what is being talked about in the classroom, often resulting in confusion or outright rejection of the science. How many teachers have been utterly bewildered when, after an enthusiastically taught unit on solutions and mixtures, their students still believe that sugar "disappears" when dissolved in water? Or following a unit on natural selection, students cling to the idea that a species can adapt to changing environments within a single generation? It happens to all of us, and our first thought is: "But I explained the science to them so clearly!"

Learning, however, is not a process of replacing your ideas with something a teacher tells you, or something you read in a book or experience in a lab. Learning is a process of actively *reconstructing and reorganizing* what you know, using a variety of resources to compare your current understandings against those of peers and against evidence in the world (you are also engaged in this process right now, using your beliefs and experiences to assess whether the ideas presented so far fit into your existing vision for teaching, or if your vision can be reorganized to accommodate these new ideas). All students come to school with remarkable experiences that should be invited into discussions—they cook, build things, ramble around playgrounds, care for pets, engage in sports, take photographs, and observe what happens in the sky or in puddles, and members of their families tinker with car engines, tell classic stories of unusual weather, and work in the garden. Teachers need to reveal and talk about these resources in the classroom to see which of them can be used, together with instructed ideas, to help students construct knowledge that they can then use to solve problems and answer questions that are meaningful to them.

This chapter describes how to start units of instruction by finding out what ideas your students use to think about the topics you are teaching. These practices, shown in figure 5.1, are the first examples of formative assessment that are interwoven into the fabric of AST. By "formative assessment," we refer to any means by which you can get information about students' ideas, abilities, or concerns that allows you to give them productive feedback or to evaluate the effectiveness of your own instruction and make modifications to better support learning. This set, which emphasizes attention to student thinking and how to influence it over time, consists of the following practices:

- activating and eliciting students' understandings and experiences that are relevant to a puzzling event or situation

FIGURE 5.1 **Core practices: Set 2**

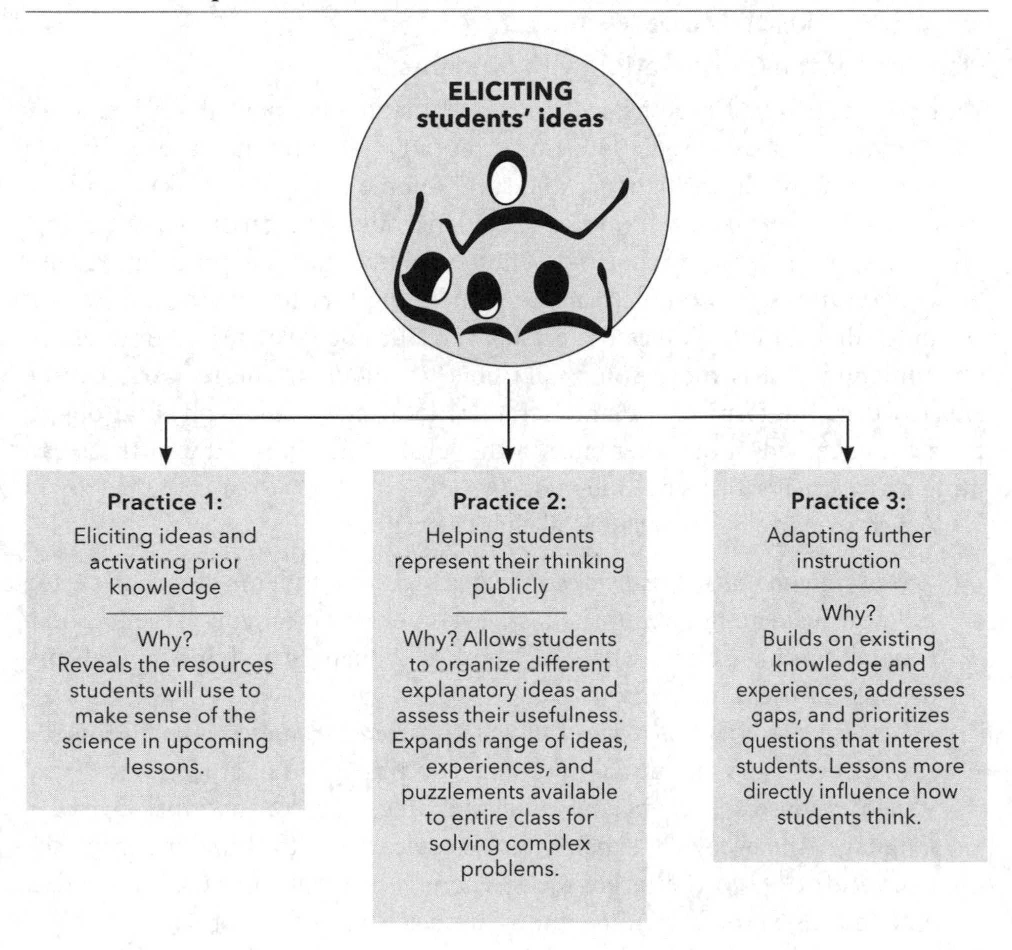

- helping students represent these ideas and partial understandings in ways that make their thinking available to their peers
- adapting upcoming instruction to take advantage of resources students bring from previous coursework or from outside the classroom, and to focus on questions they have

PRACTICE 1: ELICITING STUDENTS' IDEAS AND ACTIVATING PRIOR KNOWLEDGE

Planning for the Conversation with Students

Your goal in this first practice is to elicit what students know about just a few science ideas that are core to your unit. You can't, however, just present young learners with an interesting story and say, "Tell me what you think." You have to select a *rich scenario*, relevant to the science ideas, to get students talking. The scenario can be the anchoring event itself, but occasionally this makes initial conversations too complex and overwhelming for students at the start of the unit. Alternatively, you can select a part of the anchoring event or a related phenomenon that is more comprehensible to students. The scenario can be expressed in the form of a demonstration, local news story, physical object, puzzle, image, video, or experience in the local community. Any of these can jump-start conversations by students.

A rich scenario has two essential characteristics:

- *Accessibility to students is balanced with challenge.* Students can be expected to know just enough about the task, event, or situation you present to reasonably speculate about what they think is happening and under what conditions the event happens.
- *It has the power to reveal a wide range of ideas and hypotheses.* This means the task or event and the initial questions you pose prompt students to reveal varied resources (i.e., partial understandings, preconceptions, everyday language, and everyday experiences) they are using to think about the science and relate to the target ideas. There is no "right answer" but rather students, as a group, can imagine a number of varied hypotheses.

In table 5.1 we show samples of scenarios and initial questions that can open up students' ways of thinking to you (left column). As you examine these, ask yourself: What makes these accessible to students? Do students have a chance to at least begin hypothesizing about them? To make observations? Can students use everyday knowledge to speculate about these events and questions? In the right column are ideas a teacher might listen for in the conversation. You want to hear how students talk about and reason about these, but you are not trying to find out if students have "correct" versions of these, nor are you trying to funnel their talk toward these particular words. Rather, you are listening for

TABLE 5.1 **Scenarios, questions, and core science ideas to listen for**

SCENARIOS AND SAMPLE STARTER QUESTIONS	IDEAS TO LISTEN FOR (WHILE REMAINING OPEN TO THE FULL RANGE OF STUDENT CONTRIBUTIONS)
Curriculum topic: Plate tectonics *Scenario*: Video taken in Japanese grocery store during an earthquake *Sample questions*: ▪ What is happening to the items on the shelves at different times? ▪ What might cause the different types of shaking? ▪ Do we know how close this store was to the center of the earthquake? ▪ Where do earthquakes like this happen most often?	▪ Earthquakes result from sudden shifts in tectonic plates. ▪ These plates make up the earth's crust. ▪ Earthquakes transmit energy. ▪ This energy travels in waves. ▪ There are different kinds of waves that affect man-made structures differently.
Curriculum topic: Gas laws *Scenario*: Can crushing activity (heating water in soda can, then inverting it in cool water bath) *Sample questions*: ▪ What do you see? ▪ Does the hot water have anything to do with the can crushing? ▪ Have you seen any other examples where gases cause pressure?	▪ All gases are made up of molecules. ▪ Molecules are in constant motion. ▪ Heat energy can make molecules move faster. ▪ In contained systems, molecules bump up against the container, causing a force. ▪ There are forces exerted outside the container as well as inside.
Curriculum topic: Forces and motion *Scenario*: Skateboarder doing various tricks and jumps *Sample questions*: ▪ What are the different pushes or pulls that are acting on this person? ▪ What role do you think gravity plays in her jumps?	▪ Forces can be pushes or pulls by another object, or by magnetism or gravity, which exert constant pulls. ▪ Some objects may not move because they have balanced forces acting upon them from opposite directions. ▪ Objects in motion stay in motion unless a force acts on them.
Curriculum topic: Natural selection *Scenario*: Story about how minnow-like killifish have evolved to tolerate 8,000 times the lethal dose of toxic pollution found in East Coast estuaries *Sample task*: ▪ In small groups, after reading the current killifish story, hypothesize what you think happened with these populations before and during the era of industrial pollution in East Coast bodies of water.	▪ All organisms have structures or behaviors that help them survive. ▪ The usefulness of these structures or behaviors is applicable in particular environments. ▪ There is variance in physical characteristics within all populations. ▪ If the environment changes, the traits may no longer help the organism to survive. ▪ Organisms pass down traits to their offspring if they get a chance to reproduce.

the resources they are using to reason about these ideas. We'll give you more help on this later.

One of our most successful opening scenarios for chemistry and the gas laws involves a railroad tanker car. We use this story to elicit ideas about how gases behave in open and closed systems. The scenario is presented in both video and story form. The mystery goes something like this: A group of workers in a railway yard were asked to clean the interior of a tanker car. They opened the hatch at the top, then used long steam wands to scour the inside walls. They then closed the hatch, sealing the tanker. But there is one problem—the hatch was supposed to be left open. After a few minutes the workers began to hear unusual creaking sounds, and they immediately stepped back from the tanker. A moment later it imploded with a thunderous bang.

Over the last few years, a number of our teachers have recounted this story, shown the video, and then asked students what they think is happening (the actual types of questions and the sequences we use are in the upcoming description of the discourse itself). Each time we roll this video, the students are startled at the sudden and powerful collapse of the tanker car and they spontaneously begin to hypothesize: "The steam exploded it," "An air foot crushed it," "There was sucking pressure," "The steam melted the walls," or "It was a magic trick!" We have rarely had trouble getting them to talk. We have learned, however, to plan for specific ways that students can offer their thinking, so we can identify the different ideas they are using to make sense of this phenomenon. The following three sections share how this kind of dialogue unfolds.

Framing the Lesson

An eliciting lesson starts with framing. Framing a lesson is not the same as simply introducing what is about to happen. This is a lead-in for the conversation in which you express an authentic reason for investigating a particular set of ideas, help students put the upcoming activity into a meaningful context, and make explicit what kinds of participation will be expected of them. Here's an example of framing that one of our middle school teachers used for the unit on the imploding tanker car (figure 5.2).

This looks like a lengthy stretch of talk by the teacher, but it actually took about ninety seconds. It was time well spent because it oriented students as to why they would be studying gases from a scientific standpoint as well as an everyday standpoint, acknowledged links with the previous unit, clarified

FIGURE 5.2 **Framing by the teacher**

"So we've just finished a unit on phase changes in matter and how energy plays a role in those changes. We're now going to explore one of these states of matter more deeply–I'm talking about gasses.	**Provides context, links to previous topic**
Gasses make up the air we breathe, they keep airplanes flying, and they influence our weather. But just knowing about what makes up gasses or how they can change to other states of matter is not enough to help us explain these things we experience in our everyday lives. We need to understand how gasses behave under different conditions to make interesting things happen. So today we're going to start by talking about a strange event that happened with gasses and a railroad tanker car. I'll tell you more about that story in a moment.	**Reasons for investigating these ideas**
First, though, I want to remind you of how we participate in these discussions. The goal for us today, as a group, is to share all ideas we have about what happens in the video I'm about to show. I want to hear your thinking–and this includes any experiences you've had outside of school that you think relates to our puzzle. There isn't any correct answer for this puzzle; it's like authentic science, where a lot of different explanations are possible. There will be times when you may want to talk about someone else's ideas, and you should feel free to make those kinds of comments today, but do be mindful of our norms" (points to posters on wall).	**Expectations for participating in the intellectual work and how ideas will be treated (this part of framing can be delayed until after the video)**

the goals for the conversation, and set the expectations for participation. Does framing like this change how every student participates in the activity that follows? No, but it does increase the chances that *more* students will understand how to offer their ideas and that they'll interact with peers in the discussion.

Introducing the Scenario and Eliciting Observations

After framing, you can introduce the scenario. Some teachers begin with "I recently saw something that puzzled me," or "Let's think about this story and see what kind of sense we make of it." These are invitations for talk. As the demo, video, picture, or graph is shown, you ask students to observe closely what they see or hear going on. We *start with eliciting observations* rather than probing for explanations so all students feel like they can contribute. You should be patient about getting comments from as many students as possible, especially those that are usually quiet. Here's a passage from a ninth-grade physical science class in which the teacher is eliciting observations about the tanker car implosion. Interspersed with the dialogue is what's going on inside the teacher's head in response to what students were saying (we asked him later). Responsive teachers have to make quick decisions based not only on what they've just heard, but also on what kind of reasoning by students they want to encourage next. You'll notice the teacher using many of the discourse moves we described in previous chapters (probing, pressing, follow-ups, revoicing, wait time, etc.)

TEACHER: So here is the strange event. This happened in a railroad yard, where train cars are brought to be cleaned out. This tanker car [shows image] was being cleaned by two workers using a long wand that shoots out hot steam [shows image of wand]. They were supposed to spend a couple minutes cleaning the inside of the car and then leave the hatch open on top. But for some reason they closed it. A couple minutes after closing the hatch, they heard a creaking sound and they decided to step back from the car. All of a sudden they saw this happen [shows picture of flattened tanker after the collapse].

STUDENTS: Ohhhh. What? Cool! Is that the tanker?

TEACHER THINKS: Okay, the students are excited, but I don't want them shouting out explanations before others even know for sure what this was a picture of. I think I will show them the video now, but ask them to write a few observations by themselves.

TEACHER: I've actually got a video of this happening, and I'm going to ask you to watch and listen very carefully. I'd like you to get out your lab journal and write any observations you have—like what do you see; what do you hear? It can be anything at all, even if it seems insignificant [shows the thirty-second video, repeating it twice, then circulates around the room for about two minutes as student write].

TEACHER: Okay, does anyone want to share an observation?

MARISA: It crushed.

TEACHER THINKS: I need to follow up. I know Marisa can say more even if she's not written it.

TEACHER: How would you describe it to someone who was not there to see it?

MARISA: It looked like something stepped on it, like you step on a can. The middle caved in.

FINN: Like you step on a pop can.

TEACHER THINKS: Hmm, can I get them to relate the tanker to this everyday pop can event?

TEACHER: What does a pop can look like when you do that?

FINN: It crushes in the middle, but the ends don't; you get an empty can and just step on it, but you have to put your foot sideways and jump.

TEACHER THINKS: Finn's comment has something to do with the rigid structure of the tanker, but maybe that can be part of our reasoning about this later.

TEACHER: Does anyone want to add to what Marisa and Finn have said?

LUCAS: It imploded all at once, just kaboom. It took about a second for it to do its thing.

TEACHER THINKS: It sounds like they are saying that the explanation may be about what is happening on the outside. And the "all at once" comment may be helpful to follow up on.

TEACHER: Do you think that is important to the story?

LUCAS: I think that it only took a second for that steam to condense and change the conditions in the tanker—

TEACHER THINKS: Uh-oh, he is going to silence other students' contributions by attempting an explanation right now.

TEACHER: [interrupting] Lucas, it sounds like you have a hypothesis about what's causing this—can we come back to your idea in a couple minutes? I'd like first to get all other observations that we think might help us figure this out. Anyone?

GEVICIA: Well, you said the hatch was closed. I could kinda see that, but I have to take your word for it.

TEACHER THINKS: This has the potential to start a conversation about inside versus outside air pressure. I need her to make more of her thinking public.

TEACHER: Gevicia, do you think that makes a difference?

GEVICIA: If the hatch was closed it means nothing could get in or out; the steam could cool off in there, but nothing could go outside.

TEACHER THINKS: I am just going to wait for another student—maybe someone will pick up on this [waits twenty seconds].

SAHIL: I want to add on to that; there is no way that air can get out, like no way there can be a vacuum in the tanker because you have the same air in there. It's like sucking the juice out of a juice bag; it collapses, but you have to take something out of it. I go to the lunchroom, they sell juice there—I get a straw and poke a hole in it and pull the juice out; it starts to crumple up.

TEACHER THINKS: Hmm, the juice bag might be another good everyday event that other students can use to reason with, but this valuable example is getting lost in Sahil's long turn at talk. I could revoice just part of his comment, or I could ask a student to revoice what both Gevicia and Sahil have said.

TEACHER: I like your reasoning about comparing the tanker to the juice bag, especially about the idea that one is closed and the other one has an opening at the top. Can anyone sum up what observations Gevicia and Sahil have made, in their own words? Nathan?

NATHAN: Okay, so, there's a top on the tanker—it was not supposed to be shut, but it was. That's what Gevicia said, and Sahil said that it was weird that the tanker could get sucked in like a juice bag because it didn't have any way for air to get out or in.

TEACHER THINKS: That's actually a good enough revoicing by a student; I don't have to add on to it myself. I do, though, want to acknowledge Nathan's attempt at summing up, so I signal to the class that working on the ideas of others is valued by us.

TEACHER: I appreciate that you tried to find connections between what Gevicia said with what Sahil said—like her idea about nothing getting in or out, and then his being puzzled about the juice bag versus the tanker.

This dialogue has taken about seven minutes; usually observations can go on for a bit longer. You do want to get all relevant (and even irrelevant) observations out into the air. Don't cut this short. You can also see the range of moves the teacher has to make in order to coax students to offer their thinking, to encourage students to talk about one another's ideas, to prevent premature explanations by students, and to revoice what some students say as a way to elevate an idea, without resorting to "Hey, you are correct!" Our teachers try to maintain a "poker face" during the eliciting of observations, being nonevaluative in expression and replies about students' ideas. During this process we don't think it is necessary to write down anything publicly about the observations, because there are more important things to make lists of later in this lesson.

In the preceding conversation, some students are already wondering if their experiences with juice bags or stepping on empty soda cans can be used to understand the tanker; other students have activated their knowledge about steam condensing into water and the difference between closed versus open systems. All of these are potential resources for reasoning about the tanker. As a teacher, you can't really look inside anyone's head to tell what's being stimulated by the conversation, but you can be sure that public talk about students' everyday experiences and understandings is much more helpful to solving the puzzle than leaving those resources untapped.

There are some common mistakes to avoid in these conversations—that is, teacher moves that *discourage* participation. Here's the "shut down" list:

- initial use of scientific language that not all students have access to—for example, starting off by referencing atmospheric pressure or transfers of energy
- requests for definitions and vocabulary (e.g., "Who can tell me what kinetic molecular energy is?")
- premature attempts to get students to talk about explanations rather than observations (e.g., "What's going on that caused the tanker to crush this way?")
- efforts to steer students toward saying a particular word or phrase (that's funneling)

Any of these can signal to students that they are now "playing the game of school"—meaning that only a narrow range of responses will count as acceptable and that they should all have this event figured out by the end of today's lesson. This can diminish the discourse and alienate individuals who don't identify with science or who feel marginalized in the classroom already. Avoid this, and keep the dialogue open to everyone.

Transitioning into Talk About "Under What Conditions Would This Happen?"

After a few minutes of talking about observations, you can make a subtle but important shift to discussing hypotheses. By "hypotheses," we don't mean the strict definition used by philosophers of science; rather, we use the term less formally. We mean students' inferences about what might be happening at the unobservable level that causes an observable event or process. Students could also hypothesize about how observable conditions, unique to this anchoring event, might play a role in how it unfolded. These are not well-formed ideas by any means. We would not expect students to have coherent and elaborated explanations before you've even started the unit, but hypotheses can express possible connections between different features of the anchoring event that students feel are important to their initial attempts at explanation.

You can initiate this part of the conversation with broad prompts like "What do you think is going on to make [this event happen in this way]?" or "So how would you finish this sentence: 'It has something to do with . . . '?" Students perceive these questions as a bit more risky than just stating observations, so you will have to consider using discourse strategies that help them feel comfortable expressing their thinking. This includes you being explicit that almost anything they share will help everyone begin to build an understanding of the phenomenon, even ideas that turn out later to be inaccurate. Another way to make students feel more comfortable is to allow them to try out their thinking in pairs or small groups before offering ideas to the whole class. All three of these strategies—special verbal prompts, being explicit about sharing ideas, using partner talk—can work together to increase the chances that students will offer more of their thinking to their peers and that a wider range of students will share their reasoning.

In the case of the tanker car, the teacher followed up the observation conversation with a five-minute demo using a soda can filled with an inch of water, then heated to a boil on a hot plate. When he then inverted the can, top down, into the water bath, the can imploded with a loud pop. In the following dialogue, he asks students to hypothesize.

TEACHER: This kind of thing does not happen every day, does it? So let's have a second discussion about what kinds of conditions have to be in place for the tanker to collapse, or the can to collapse.
STUDENTS: [Silence]

TEACHER THINKS: I wonder if I asked too vague of a question, or maybe they don't want to venture a guess. I'll try the "turn-and-talk" strategy and ask a more specific question.

TEACHER: Okay, I'd like you to turn to your neighbor and talk for about thirty seconds. The question is a fill-in-the-blank and there could be a lot of possible responses: "The tanker will only collapse if . . ." and then you fill it in [teacher circulates and listens to what different groups are talking about, then asks one pair of girls to be prepared to share].
TEACHER: Can we offer any ideas now?
KAYLA: We said that the tanker or the can has to be hot; otherwise, the collapse won't happen.

TEACHER THINKS: I'll see if I can get them to reason out loud; they may have a reason for saying that.

TEACHER: What makes you say that?
KAYLA: Because both had super-hot water or steam in them right before it happened, and if you don't heat the water, neither one is going to crush.

TEACHER THINKS: Instead of me pressing her to say more right now, I'll get more students involved in the conversation. But I'll ask them to connect to Kayla's statement if possible so they get practice commenting on each other's ideas.

TEACHER: Anyone want to add on to that?
TAYLOR: Something has to happen on the inside for it to implode. Like condensation or cooling off—the tanker cooled slowly and the can was chilled, it was like, chilled really fast.

TEACHER THINKS: Need to ask Taylor to link it to previous comment.

KAI: The lid was on; the lid has to be on. The lid stops the air on the inside from moving to the outside. Nothing goes in, nothing comes out.

TEACHER THINKS: Oops, too late, but glad I don't have to take every other turn at talk. Maybe I can now get Taylor to comment on another student's idea.

TEACHER: So, Taylor, does Kai's claim that the lid has to be closed have anything to do with your idea about condensation and cooling off?

TAYLOR: Well, if the steam can escape out into the air, then it's not going to condense inside the tanker, which causes certain forces, and then it won't implode. That has to be right because you said it was a mistake to shut the lid to the tanker—they, uh, made a mistake, so they normally must leave the lid open and nothing happens, but . . . [trails off]

TEACHER THINKS: *I need to sum up here so the other students don't get too lost in all the various ideas. I also need to have my quiet students participate. I can call on the two students who had been talking during the partner conversations about "outside forces" and "inside forces."*

TEACHER: Okay, so we have some hypotheses about condensation, cooling off, and the lid keeping something inside the tanker, or making some process happen inside the tanker that normally would not happen. Tony, you and Addy were talking about forces in the tanker; can you share more about that?

TONY: So we said the steam condenses in the tanker and then it takes up less space, like we talked about last week in the phase change unit. The molecules slow down. When it takes up less space there's like a vacuum and it creates a suction force, and then it collapses. We thought there was a vacuum force.

ADDY: Yeah, so gravity can take over and crush it when there is less air space inside the tanker. There is less pressure on the inside than on the outside.

TEACHER THINKS: *Tony and Addy are adding a couple new ideas that have the potential to help others reason about the tanker. I'll have other students revoice them.*

TEACHER: So I am hearing some additions to our list of hypotheses; can anyone sum up what Tony and Addy said made a difference? Yes, Finn?

FINN: Condensed steam takes up less space, the molecules move slower, the less space makes a suction force, the gravity is enough then to make the tanker or the pop can implode. But I disagree with the last part—I think it is air pressure on the outside that makes it crush.

TEACHER THINKS: *Finn has a lot of the pieces for an explanation, but they are loosely connected. "Suction" theories are misconceptions, but no time to address them now.*

TEACHER: Did the air pressure on the outside change?

FINN: Well, maybe not.

TEACHER THINKS: I could keep going with this line of questioning, but some students will tire and tune out. I'll move on to a more active part of the lesson in which I can see ideas from everyone.

TEACHER: Okay, Finn, but we'll still keep that outside air pressure idea as something we can test out later in the unit. Thanks for helping us remember the different ideas that came up, and adding your disagreement about where the pressure is coming from. We have ideas [writes these on the whiteboard] about molecules moving, steam condensing, suction forces, gravity, air pressure, and the lid being closed. Right now it sounds like we are unsure about whether the cause of the collapse is because of what's happening inside the tanker, or maybe outside the tanker, or both. I'd like to see what everyone is thinking about this, so right now I want you to work with your partners, take one of these sticky-notes, and write on it "inside," "outside," or "both." Then below that, write which one or ones of these ideas on the whiteboard you think may be most important to explaining the tanker.

In these exchanges, the teacher has tried to open up the conversation by using probing questions, wait time, turn-and-talks, revoicing, follow-up questions, and having students comment on one another's ideas. The resulting talk has surfaced different kinds of reasoning by students. Many ideas—some that are relevant, and some that are not relevant or even consistent with science—have been offered. None have really been unpacked thoroughly or linked tightly with other ideas, but the goal here has been to reveal as many of the reasoning resources that students are using as possible. Even the student's analogy to the straw in the juice bag is a resource to reason with.

As a group, students have had ideas *activated* that they might never have considered if they were working alone. They've heard peers talk about outside air pressure, open versus closed systems, the effects of gravity, cooling and condensation tied to slowing of molecular movement, and vacuums. These aren't perfectly understood by all students, and no one has a grasp on how they all work together, but most students likely feel that these ideas can be tools to work with as they build their explanations.

Providing "Just Enough" Information to Elicit Ideas

When eliciting ideas for some kinds of subject matter, you'll need to first give students information in order for them to be able to reason about the scenario. For example, in the high school biology unit about the reintroduction of wolves to Yellowstone National Park that we mentioned earlier in the book, we wanted to elicit students' thinking about how different populations of plants and animals could interact with one another in an ecosystem. But even though we believed students had a lot of prior knowledge about ecosystems, we also knew that their lack of familiarity with the particular species of plants and animals in Yellowstone would keep them from revealing what they knew about interactions. So when it came time for the eliciting conversation, we placed students in groups of three and distributed to them several "species cards," each listing a different member of the Yellowstone ecosystem. On each card was the name of a species, its habitat, food sources, and life cycle details.

After they read the information on the cards, we heard students venturing guesses about how populations might fluctuate based on the availability of resources, and how some species might indirectly influence the success of another species. Only after they had done this brief round of reading and small-group discussion did the whole-class conversation begin. In giving them this information, we weren't aiming to teach them how long a beaver lived or the mating habits of mule deer; rather, we wanted to hear how they reasoned about these animals' roles in the ecosystem.

Here's another example of providing minimal information. In chemistry classes, our teachers often use fireworks as anchoring events to help students learn about atomic structure and electron shells (when electrons in some types of atoms receive energy, they jump temporarily to an outer shell; when they jump back down they release energy in the form of photons, which our eyes detect as different-colored bursts of light). Students think fireworks videos are interesting, but they can't offer any ideas about what is causing the colors. We soon learned that if they were simply given schematics of typical fireworks devices and information about what is inside them, they could begin to hypothesize about what might be going on at unobservable levels to produce the dramatic displays. Regardless of the topic or grade level, when we plan to elicit students' ideas at the start of a new unit, we always wrestle with the question: What is *just enough* information for them to reason about, without giving away the science itself?

PRACTICE 2: HELPING STUDENTS REPRESENT THEIR THINKING PUBLICLY

Up to this point there has been a mix of whole-class and partner conversations, lasting anywhere from fifteen to forty-five minutes. Now is the time to help students make a record of their initial ideas. This is an authentic scientific practice; students make explicit what they currently believe about the science behind the scenario. More importantly, they will revisit these representations in the middle of the unit and at the end of the unit to make changes.

Why is this practice useful? First, it allows students to take loosely connected ideas and begin to clarify if and how they are related to one another. Second, because these inscriptions (drawing and/or writing) will become public, students can see what resources their peers are reasoning with and learn how different ideas can be represented. Third, you can rapidly survey the ideas students are thinking with, in order to see what science they already know, what interests them, and where their gaps might be.

Option 1: Create Small-Group Models of the Scientific Phenomenon

Have students work in pairs (preferred) or threes to create *small-group models* of the phenomenon in the scenario. This is the most popular option among our teachers. Students are asked to show what may be going on "before, during, and after" the phenomenon happens. They are encouraged to draw out both what is observable and what is unobservable. In the case of the tanker car, everything happens in a matter of minutes. In the case of the wolves of Yellowstone, the phenomenon unfolds over decades. If you choose this modeling option, you should give the students ample time to do the work because they are not just recording their existing ideas, they are thinking and revising a lot of mini-theories as they are drawing.

The full eliciting lesson, then, nearly always stretches out over two class periods or one block period. After everyone's finished with their models, our teachers have students do a gallery walk and ask them to look closely at what ideas are included in the drawings of others. The next chapters focus on the scientific practice of modeling, and we share much more about how to do this with students. Figure 5.3 is an example of an initial model of the tanker, drawn by two eighth-grade students.

Option 2: Create a List of Students' Hypotheses

An alternative to modeling is creating a class list of students' hypotheses. To get started, you can put up two sentence frames on the board, so students know

FIGURE 5.3 **Initial tanker model by two eighth graders**

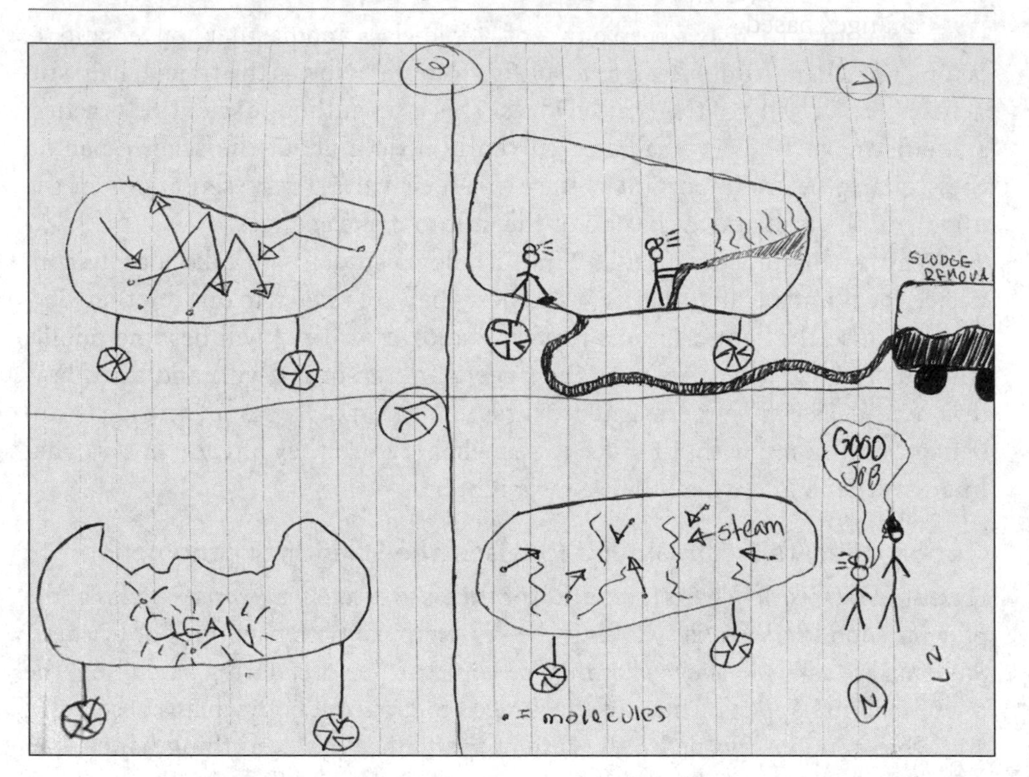

how to participate. One is: "We think [the phenomenon] has something to do with _____." The other is "We think [the phenomenon] happens the way it does because _____." The first frame is easier for students to contribute to because it does not require a ready-made causal story; they only have to suggest a condition that seems to be necessary for the event to occur.

You moderate the construction of this list not by writing everything students say, but rather by asking the class, as the list is added to, "Is this hypothesis different from the others? How? Can we combine your hypothesis with theirs? Do you mean . . . ? What are some things we are not sure of?" Only after having students talk to each other about whether hypotheses are different, or how they are related, might you add one to the public list. You want between

four and six hypotheses. The whole class will revisit this list later in the unit to make changes based on new evidence and information.

Figure 5.4 shows a list of hypotheses, generated in a sophomore biology classroom, focused on cells as units of life. The anchoring event was the teacher's relative being diagnosed with cancer. The essential question was "Why did it spread and spread so fast?" The figure includes hypotheses generated by the teacher's first three class sections.

On the sticky notes below each hypothesis are the names of the students who believe that statement is closest to the actual explanation; everyone in the class had to participate in generating the hypotheses and in casting a vote. Many of these statements are examples of partial understandings—for example, #1, #2, and #5. At least a couple of these make even the most experienced of teachers wonder if they are valid—for example, #6 and #7. Does cancer weaken the immune system? Does exposure to toxins cause preexisting cancer cells to multiply more rapidly? Most educated adults *don't* know whether these are true or not, and the teacher in this case readily admitted this, then joined her students in learning more about cancer. In the process, she modeled "not knowing but being systematic about finding out" to her students. As the unit progressed, the teacher asked students if they thought (1) some hypotheses should be discarded, (2) some hypotheses should be joined with others, or (3) new hypotheses should be added.

Revised Thinking

In the fifth-grade class studying sound that we've described previously, students produced a class list of "theories" for why the glass broke when the singer produced a certain musical tone (see figure 5.5, a variation on the list of hypotheses). In the space below each statement, students commented throughout the unit, in words and pictures, about what activities or readings support each of these theories or make them reconsider a theory's inclusion in the list.

Both options for publicly representing thinking—creating the models and the lists of hypotheses—can be used as community tools for further intellectual work. Either of them can be developed, added to, subtracted from, or reorganized by students as the unit progresses. No matter which kind of representation you create with students, you can ask at the very end of class: "What questions do we have now about the phenomenon?" and "What kinds of information or experiences might we need to learn more?"

FIGURE 5.4 **Hypotheses for origins of cancer, generated by multiple class sections**

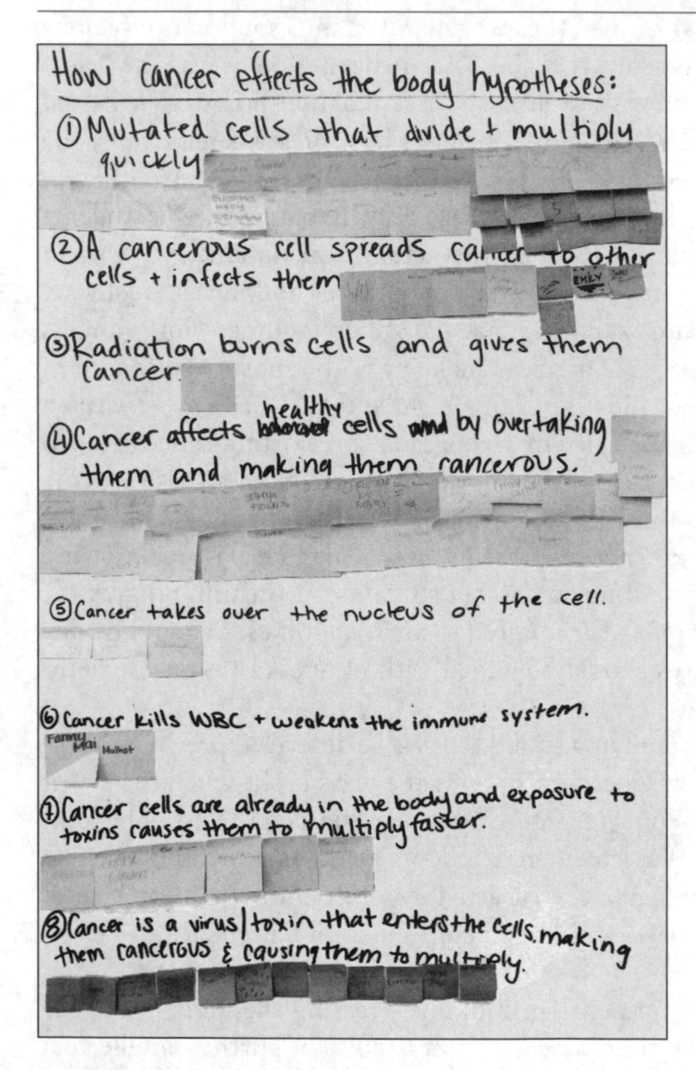

1. Mutated cells that divide and multiply quickly.
2. A cancerous cell spreads cancer to other cells and infects them.
3. Radiation burns cells and gives them cancer.
4. Cancer affects healthy cells by overtaking them and making them cancerous.
5. Cancer takes over the nucleus of the cell.
6. Cancer kills white blood cells and weakens the immune system.
7. Cancer cells are already in the body and exposure to toxins causes them to multiply faster.
8. Cancer is a virus/toxin that enters the cells, making them cancerous and causing them to multiply.

FIGURE 5.5 **Fifth graders' initial hypotheses about how a singer breaks a glass with his voice**

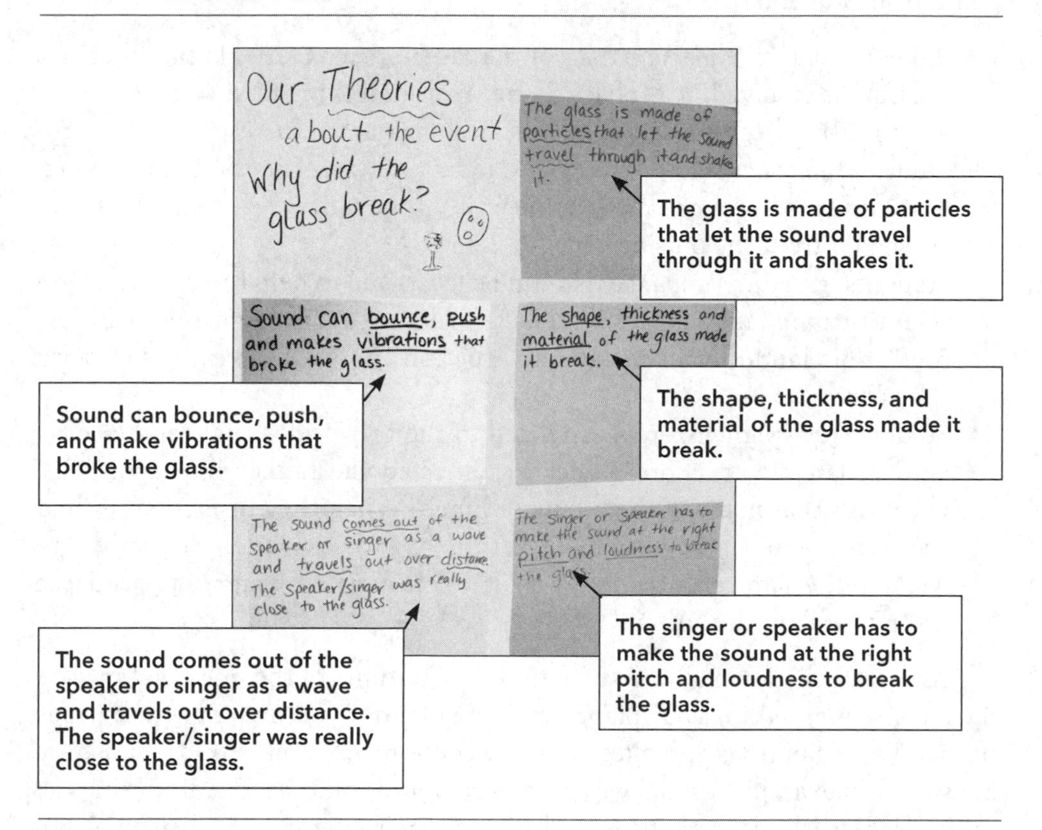

PRACTICE 3: ADAPTING FURTHER INSTRUCTION

This final practice in the eliciting set is something you do after the school day is over. Take a few minutes to sketch out what students talked about in terms of partial understandings, alternative conceptions, everyday vocabulary, and even ways of arguing about the initial puzzle or event. You should also make note of the everyday experiences or events that your students used to talk about the scenario. You then have to exercise judgment about how to use some of these ideas as resources in upcoming lessons, or make plans to address gaps or preconceptions in students' thinking.

Here are some examples from the tanker lesson about what the teacher heard from students:

- Students' *partial understandings* of the target ideas. Students said that "air pressure is involved," or "phase change had something to do with it." These are mini-theories that can eventually become part of full explanations.
- Students' *alternative understandings* about the target ideas. Students said that "a sucking pressure drew in the walls of the tanker," or that "the steam crushed it from the outside."
- Students' *everyday language* that can be leveraged to help them understand scientific language related to the target idea. Students referred to an "air foot" that landed on the tanker. "Suction" is another everyday word they used.
- Students' *everyday experiences* related to the target idea that can be leveraged in later instruction. Students connected the tanker implosion with juice bags that many of them buy for lunch. When they insert a straw and suck out the juice, they can easily get the bag to "implode." Similarly, students in the same class said they could make a plastic water bottle collapse just by "pulling the air out."

As you distill down all the students' contributions to the ones that may be most consequential for your planning, it may help if you map out the various partial understandings and alternative conceptions on a piece of paper. Figure 5.6 shows an example of what we heard from middle school students during the initial "can-crushing" activity. No teacher has the time to create such an elaborate diagram, but we wanted to represent the kinds of logic that can be used to adapt upcoming lessons.

We try to note how prevalent these ideas or experiences are among our students, how enthusiastically they expressed particular ideas, and the relevance of students' resources to the science itself (appendix E shows a tool we use to do this, the Rapid Survey of Student Thinking). Just because a student used a particular idea or experience in talking about the anchoring event does not mean that you should make changes to your upcoming lessons; however, most of our teachers uncover a few conceptions they never expected in students' conversations—something that requires a modification of what they'd planned.

What does modification look like? Well, *it's not throwing out your curriculum*. It often involves adding a lesson, rearranging the order of lessons, or

FIGURE 5.6 **Model of how upcoming lessons can be adapted, based on students' initial ideas**

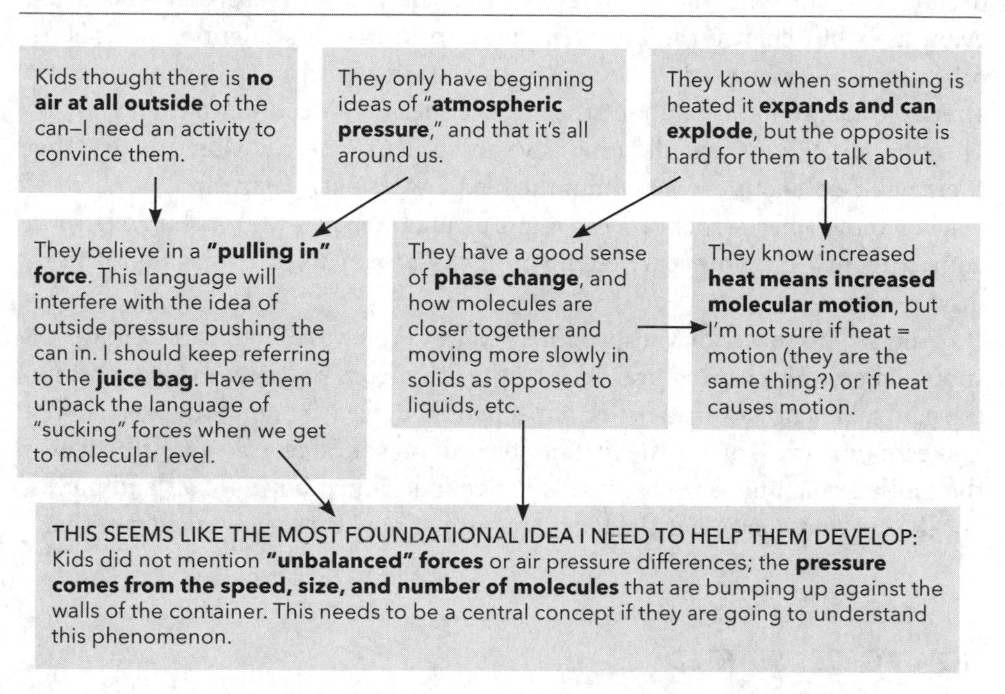

integrating new phenomena into the lessons. In the first lesson of the aforementioned fifth-grade unit on sound energy, the teacher found that her students did not understand the difference between sound moving from one place to another and air itself moving from one place to another. The teacher felt this unexpected and fundamental confusion had to be addressed early, so she created a lesson to help her students understand "what's moving" when sound travels versus when the wind blows. In another example, while doing the first lesson of a middle school unit on electricity, one of our teachers found that his students were not interested in generic batteries, wires, and bulbs, but were eager to understand a puzzle posed by one of their classmates: "If I have to stay overnight at my uncle's house, and he doesn't have a cell phone charger, can I charge my phone with his phone?" That became the essential question of the unit. All the big science ideas were explored as the teacher had originally planned, but in a context that motivated students and piqued their curiosity.

In some cases, when you find out about your students' thinking, you may decide to modify the anchoring event itself, or you may keep the anchoring event as is but change the essential question to match students' interests. In other cases, you may add a lesson or two, or, if you find that your students already understand a lot about the science ideas, you can cut out an activity. These modifications are all responsive to students' needs. Don't expect that there will be obvious choices about the next few lessons. There are no "correct" teacher moves here—this process requires judgment as well as knowledge of your students and your curriculum to make a principled choice about moving forward.

And as you move forward, you may want to refer to a map we've created; it shows where this eliciting practice and the other core practices are used during the course of a unit of instruction. In appendix B, we lay out all the lessons that make up two units: one is the fifth-grade unit on sound energy and the other is the tenth-grade unit on ecosystems. We've mentioned both in bits and pieces, but the appendix provides the bigger picture of how these lessons fit together.

HOW TO GET STARTED

In your first attempts at eliciting students' ideas, we recommend that you look back at the two chapters on discourse and identify the moves you think would be helpful to get your students talking. Here's one tip: decide ahead of time what you'll do when you do not get any responses from students. This happens a lot, especially when students aren't familiar with a kind of teaching that does not seek right answers to fact-based questions. They'll be a bit suspicious and reluctant at first, but they'll soon learn the rules of this eliciting game. Your careful framing of the lesson can also help put them at ease. If it's your first time framing a lesson, try scripting it out before class starts.

We strongly recommend recording video of these lessons. You can then see which of your talk habits get students to say more about their reasoning, and how these lead to small successes or new challenges. When you're watching the video, two good indicators of success will be how *long* the responses are by your students (are they more than just a few words?) and how many *different students* contributed during the class period. These are just rough metrics that give you indications about how productive the class conversations were, but they can also reflect rigor (chances to think out loud) and equity (who gets to participate).

Here is another outcome to assess if you have students draw out small-group models: check how different these initial representations are from one another. If everyone's models look the same, it could be that there is only one reasonable explanation and that the scenario is not complex or puzzling enough. On the other hand, if the models are empty, perhaps students needed a little information about the scenario or anchoring event to theorize what might be going on. In the next two chapters, we talk more about how to get started with designing modeling opportunities for your students. You may want to read the modeling chapters before teaching the kinds of lessons we have just described.

SIX

MODELING, PART 1

Making Thinking Visible Through Models

WE CAN LEARN a lot about scientific modeling by comparing examples from two classrooms at opposite ends of the K–12 spectrum—high school advanced chemistry and kindergarten. We'll start with the chemistry students and a unit on thermodynamics. Early in this sequence of lessons, the teacher began class by brewing a fresh pot of coffee at the front of the room. On a nearby table he had placed Styrofoam cups, a container of ice, and thermometers. Students were asked to pour the coffee into the cups, then use the available materials to determine the temperature of the coffee without measuring it directly. The teacher wanted students to end up with accurate estimates, but beyond that, he wanted them to explain, through modeling, what was happening in their experimental setups that allowed them to make these estimates and trace the movement of heat in the system. In other words, he was requiring a "why" explanation. Before the hands-on work commenced, students sketched out tentative models; they talked, argued about what to show and how, visited other groups, and borrowed lots of resources (ideas, language) from peers to reconstruct their first-draft drawings. Figure 6.1 shows how students in one group ended up using multiple representations within their model to reason about the transfer of heat; they incorporated equations, pictures of molecular kinetics, and written text to make sense of what might be happening in the cups. Part of the challenge was that all these symbols, words, and numbers had to tell a

FIGURE 6.1 **AP Chemistry model by pair of students on heat transfer**

Using calorimetric techniques, determine the temperature of the water in the coffee urn. Standard tub materials are available, but your thermometer can only read to 50.0°C. Ice must be used in your process.

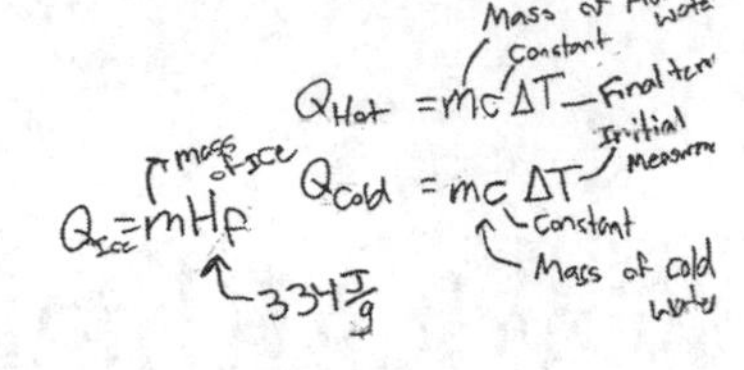

As a part of your planning, create a visual representation of your physical process for the lab.

- Use force vectors to describe the relative energy that is present in each aspect of your experimental design.
- Include an indication of which direction energy is moving during the calorimetric process.
- Show pertinent equations that may be usseful in solving the lab problem.

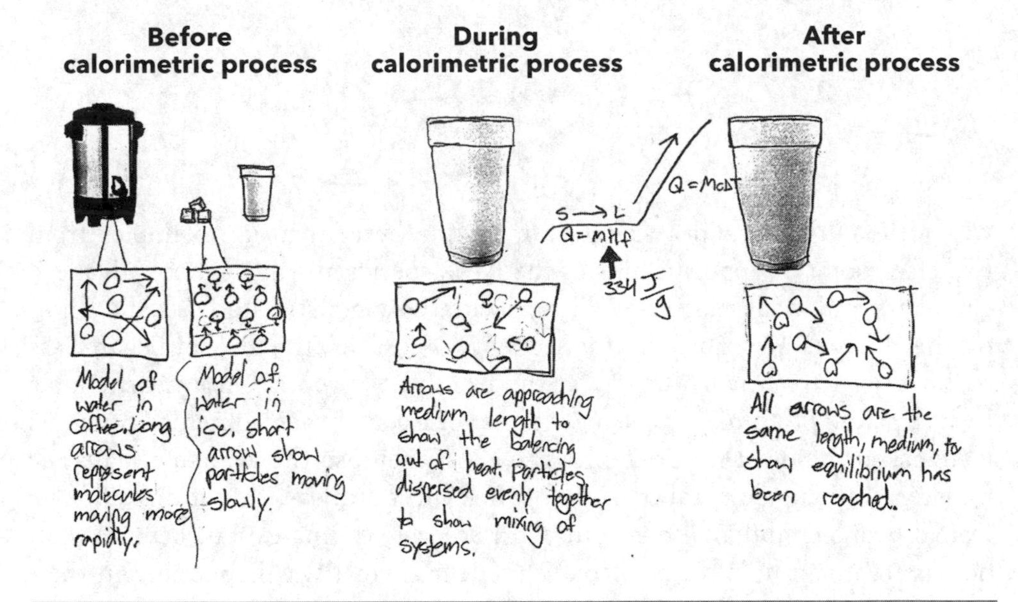

coherent story, together. By the end of this sequence of lessons, no two of the dozen or so small-group models were alike, yet most of them embodied scientifically feasible explanations. All of this thinking and drawing appears remarkably generative, but is modeling something that only older, more advanced students can do?

In the kindergarten classroom, students were busy explaining how someone little could bump someone big off the end of a playground slide. They started this force and motion unit by going out to the playground, taking turns

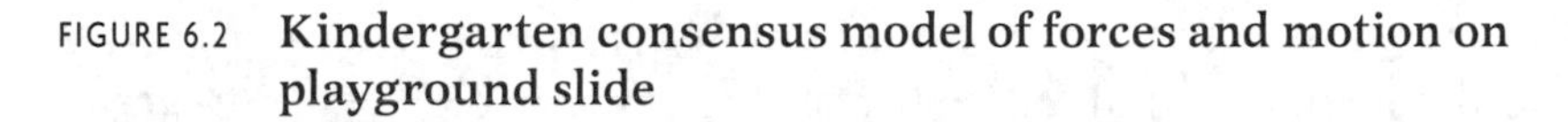

FIGURE 6.2 **Kindergarten consensus model of forces and motion on playground slide**

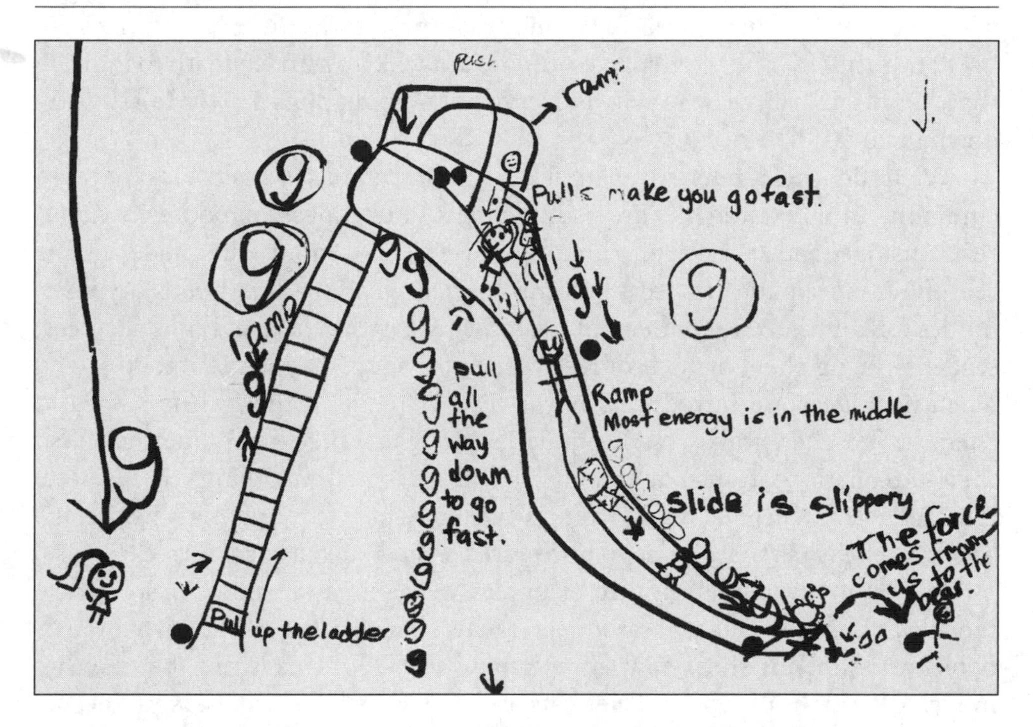

on the slide and colliding with a teddy bear placed at the bottom. Figure 6.2 shows a consensus model that the teacher and students started that day, then added to over the next three weeks. What began as a simple sketch of the slide, with a time-lapse rendering of a person sailing down, was annotated with many new ideas over the course of the unit. Importantly, this model was not just an outcome of classroom conversations; it had generated debates among students that led to changes in their thinking.

Early in the unit, several boys and girls volunteered to come to the front of the classroom and draw "pushes and pulls" in places on the model that even the teacher had not considered, occasionally demonstrating, for example, how one might hoist themselves up the first rung of the ladder. A few days later, after a balls-and-ramps activity, some students convinced the teacher that "the forces come from us to the bear" at the bottom of the slide, while others added that "some of the energy may be in your shoes." Still later, when the idea of gravity was introduced during a read-aloud, many of the five-year-olds began to argue why and how it should appear in the model. At first it was inscribed on the slide itself; this caused other students to propose a stack of *g*'s underneath to indicate that, as one girl claimed, "gravity is everywhere" and it pulls on you "all the way down to go fast."

What do the kindergarten and advanced chemistry scenarios have in common? In both, students are using models to test ideas, hypothesize about relationships, and revise their thinking as they learn more. The models make reasoning more public to peers and the teacher, which in turn opens up possibilities for continued refinement of the drawings by the class. In both instances, students bring their own resources to help produce the initial models, and when new ideas are introduced by the teacher (equilibrium by the advanced chemistry teacher, gravity by the kindergarten teacher), the combinations push their explanations in new directions. Our point is that modeling can facilitate this valuable intellectual work, from early elementary through high school. It's one of the reasons that this scientific practice plays such a central role in AST.

As you might guess, the two examples we've cited are not typical of how models are used in most classrooms. Teachers more often use models in textbooks or online to "show how something works." But this is not the role that models play in the discipline. Scientists use models to make unobservable events and processes visible so members of their community can pose new questions for investigation that may result in changes to the model, and to construct better explanations for how the natural world operates. Because they are so central to disciplinary work, drawings (and redrawings) of models are everywhere—scrawled across whiteboards in laboratories, recorded in the pages of field journals, and displayed to audiences at conference presentations. Without models, much of the scientific enterprise would come to a standstill.

As we've seen, students are more than capable of illustrating their thinking about complex phenomena through drawing and writing, and just as in science communities, their models can also be used to generate questions, deliberate about what information they still need, and revise explanations together. In classrooms, as in science, modeling is not a one-and-done event ("Let's do a lab activity and make our model, and then we'll be finished"). It is a dynamic process that unfolds over time as students gradually *change their understandings* of an event or process being studied. As students learn more, they may add new

Revised thinking

cause-and-effect relationships, discard some parts of a model, or realize that labels for what is happening need to reflect new scientific language. To put this another way, much of the teaching activity during a unit is directed toward the goal of helping students understand how they can reorganize or reconstruct what they know, by working on and with models over time.

In this chapter we explore a number of questions that teachers often ask: *What counts as a model? Are the models that scientists use different from the ones my students should be developing? What does the process of modeling look like in the classroom? How does modeling help students learn?* We address the basics of modeling and show how you can launch this process in the first few days of a unit. Then, chapter 7 takes you a step further by describing how models can be revised over time and how they support learning across the trajectory of a unit.

WHAT ARE MODELS?

A scientific model is a representation of a system (such as the human respiratory system, the solar system, or a system of electrical circuits) or a phenomenon (such as the changing seasons, the oxidation of metals, or mammals maintaining their body temperature). These representations can take the form of drawings, diagrams, flow charts, equations, graphs, computer simulations, or even physical replicas such as a tabletop model of a watershed. Some types of models *describe* parts of a system or phenomenon, and other models *explain* how or why something happens. We focus on the latter, explanatory models, in these chapters.

Models usually include only features that are important for understanding the system or phenomenon they represent, and leave extraneous information out. For example, in modeling a system of pulleys and weights, we might draw the pulleys, the strings or ropes, the weight, perhaps who was exerting a lifting force, and the forces themselves. What might be left out are trivial surface features of the object being lifted and the details of the person or thing doing the lifting. Neither of those would help us explain how the pulley worked, or help us predict whether the pulley could lift a particular weight.

Scientific models have to change in response to new forms of evidence about the world. Scientists refine models so that they can more accurately explain a wider range of events and processes. The modern models of DNA replication help us explain some of the same patterns of inheritance that Mendelian models of genetics did over a century ago, but the most current models also suggest why, for example, traits appear to be "switched on or off" in response to environmental conditions outside the organism. An improved scientific model is usually consistent with both new and previously established scientific evidence.

Modeling works most closely with another scientific practice—explanation. In fact, a well-labeled model can be a form of explanation. Written explanations, however, include details that cannot be expressed easily in models. Models, in turn, portray key parts of causal stories that are often difficult to put into words. Thus, these two practices operate hand-in-hand, and both are at the heart of scientific work. Scientists use modeling in concert with other practices as well. As you can see in figure 6.3, some of these practices *inform the development* of models, such as analyzing data or communicating findings; other practices are *motivated by ideas proposed* in models, such as asking new questions or engaging in argument; and some practices, such as designing and carrying out investigations, *both* inform and are motivated by ideas in models. Note: please don't confuse teaching practices with scientific practices; these are different kinds of activity, although you use teaching practices to help your students engage with scientific practices.

WHAT TYPES OF MODELS WORK WELL FOR MODELING IN THE CLASSROOM?

How Models Are Commonly Used

Teachers have always used models in the classroom, but the most frequent purpose is to illustrate accepted science ideas. While this can be part of purposeful instruction, it is not modeling. Even when teachers ask their learners to draw out their understandings in pictures or diagrams, such displays are typically disconnected from knowledge-building activities—students simply "posterize" (create posters of) science ideas that are fully described in the curriculum. We've all seen hallways in schools covered with nearly identical representations of trees going through photosynthesis, cells going through their mitotic phases, the water cycle, and the rock cycle. One could say this is "making models," but unlike modeling, these experiences don't support learning very well, in part because they don't require students to solve problems they perceive as meaningful, to develop ideas, or to test connections among ideas. Students are reproducing the end products of other people's thinking rather than doing the thinking themselves.

FIGURE 6.3 **How scientific practices inform models and explanations, or are motivated by them**

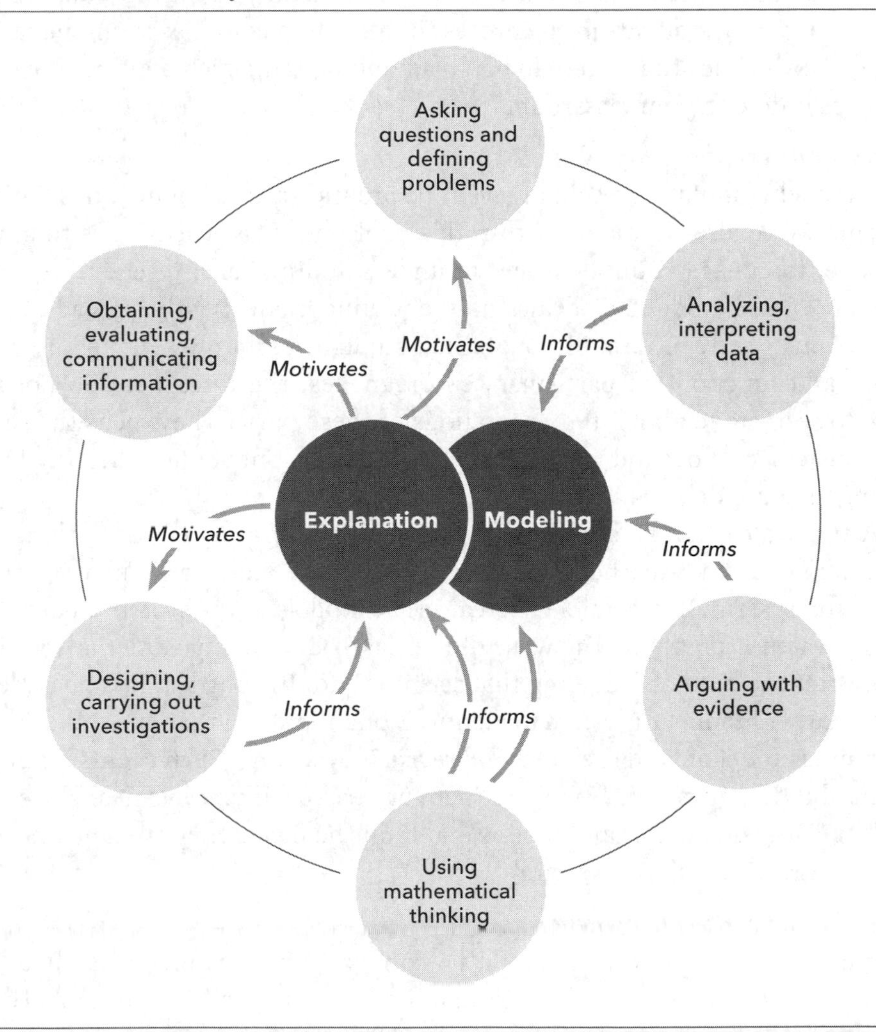

It is also increasingly common to have students use whiteboards to record data from lab activities and to share these out in class. The graphs and charts are *models of data* rather than *models of phenomena*, and the practice is indeed helpful for student learning. It is not, however, the kind of modeling practice that is explicitly used for building knowledge about an event over time.

Principles for Setting Up Modeling Experiences with Students

To help you make the transition from using models only in the aforementioned ways to engaging students in the process of modeling complex phenomena, we share six principles that can help you plan and be strategic about how science practices unfold in your classroom.

Represent an event or process

First, models should represent an event or process, rather than things. This is a theme we've already mentioned in this book, but it is worth revisiting. For example, to engage young learners in understanding cells, teachers we work with have asked students to draw and then refine models of the spread of cancer in human body tissues. Although their students certainly need to know the names and functions of particular cell organelles, these teachers do not ask them to recreate textbook representations of these parts. They focus their students instead on how and why these cell structures contribute to healthy functioning or cause disease.

As another example about shifting modeling from being about things to being about events, teachers often have their students represent the earth-moon-sun system. It is possible to create scale models of the system's parts and arrange them properly on the walls of the classroom—many students do this—but it's not the kind of modeling that scientists do. In contrast, it is possible to use the earth-moon-sun system to identify a phenomenon that one could create a dynamic model of, and then test and revise it over time. Such events might be captured in the questions "Why are there no seasons if you live near the equator?" or "Why do planets and moons maintain the orbits they currently have?" or "Why are solar eclipses so rare?"

Select a context-rich phenomenon

The phenomenon you select should be context-rich, meaning it is about an event that happens in a specific place and time, under specific conditions. These "specifics" are precisely what make the models interesting to students. As we mentioned earlier in the book, a unit on ecosystems used by several of our teachers asks students to wrestle with the question of why killer whale populations in Puget Sound are declining. Students are given information about a specific community of whales, the changing numbers of adults versus juveniles in that community, and the number of males versus females. This is not a generic scenario; it is about a real group of mammals that lives in a unique environment. This environment includes a dynamic web of other organisms, seasonal changes in the habitat, and several sources of pollution. Having to piece together how these contextual features might affect the decline of the orcas makes it much more challenging for students to construct their models and allows for diversity in how students represent their thinking. Both of these conditions are highly conducive to learning.

Make models pictorial

Students can represent ideas more effectively if their working models are pictorial, meaning there is some visual resemblance between the real conditions being modeled and what gets put on paper. In the fifth-grade unit on sound energy we've discussed, where the anchoring event was the singer shattering a glass with the energy from his voice, students' models included a picture of the singer's head, the glass, and representations of sound itself in the form of waves, ripples, or particles. This phenomenon is fairly straightforward in terms of how it translates to a drawing. Other events and processes, however, can be more complicated to represent pictorially. In middle school units on earthquakes, our teachers have found it helpful to have students sketch both a top-down view of the land being affected and a "cutaway" or side perspective of the earth. Some phenomena are more amenable to pictorial representations than others, but we've identified strategies that could make modeling feasible, regardless of the topic. For example, because students can struggle with the drawing before they even get to showing their understanding of the science, we've found it helpful to provide them with model templates. We address this in an upcoming section of this chapter.

Some forms of models are less functional for classroom modeling; these include computer simulations, graphs, equations, or physical replicas. Although they are powerful tools to help students reason about ideas (we use them regularly in AST), they are not the types of models that students can readily test, evaluate, and revise over time.

Include both observable and unobservable features

The fourth characteristic of models for classroom modeling is that the representations should include both observable and unobservable features in the world. Features included in an explanatory model might be unobservable because they are inaccessible to our senses or to measuring instruments (e.g., the layers of the earth or hormonal reactions in the body), are too small (e.g., sound wave transmission, chemical reactions at the molecular level), happen on a vast scale (e.g., the blocking of the sun's light during an eclipse), happen over long periods of time (e.g., stellar life cycles, evolution, continental drift), or are conceptual (e.g., selective pressure in ecosystems, unbalanced forces). Representing these features takes creativity and some agreement among students about how marks on paper will stand in for fundamental science concepts that are abstract, yet shape everyday events we see in the world.

Simply put, explanatory models in science use unobservable features, events, processes, and structures to explain what we can observe. In our sound example, students can observe (and draw) the singer holding the glass close to his face, hear the high-pitched tone, and see the glass break. Students are then asked, "What's going on that we can't see?" They hypothesize about the unobservable—the sound waves, the air itself, vibrations in the glass, and even the singer's lungs. This is what is meant by *theory*; it is using the unobservable to help explain what we can observe (meaning phenomena or patterns in data). The relationship is two-way because what we observe with our senses or with instruments is used as evidence to create a theory or an explanatory storyline about the unobservable. This relationship between observation and theory is a powerful disciplinary idea and it holds true for all fields within biology, chemistry, physics, and the earth and space sciences.

Include conditions before, during, and after the event

Modeling the before, during, and after seems too obvious to point out, but all events and processes take place over time. There is always a set of conditions that exist before something happens, while it happens, and then after it happens. This is important for learning because students can produce much richer models and written explanations if they include these different points in time. For example, in a high school unit on how the human body uses homeostasis to maintain its internal conditions, one of our colleagues used the situation we touched on in chapter 3, of a runner who suffered a mild heat stroke during a summer road race. Using the template shown in figure 6.4, the teacher asked her students to model what was going on in this runner's body before the race got under way, then during the heat stroke, then as the runner was recovering. (Notice in figure 6.4 the creative use of the empty graph running across the top,

FIGURE 6.4 **Before-during-after model template for homeostasis during exercise**

Name: ______________________________

What you'll do: You have three stages of a healthhy runner completing a race below. Before he starts running, during is when he is running, and after is when he has stopped running. In the case of our runners in the 5K High School boys race with an external temperature of about 100 degrees Fahrenheit, these images represent healthy rinners (where homeostasis is working properly). Using the template below, draw what happens at **the unobservable level and the observable level.** Use the zoom-ins to indicate what is happening at the level of the skin, brain, and muscle. When you draw something, **label it with a word or two.** Please **use the drawing convention we talked about in class**—a curvy arrow to indicate heat where the length of the arrow corresponds with amount of heat. Also, indicate the runner's body temperature as it changes over the three plases with the graph on top (hint: remember our activity with our own temperature graphs?). Do not worry about specific numbers here, just focus on general trends.

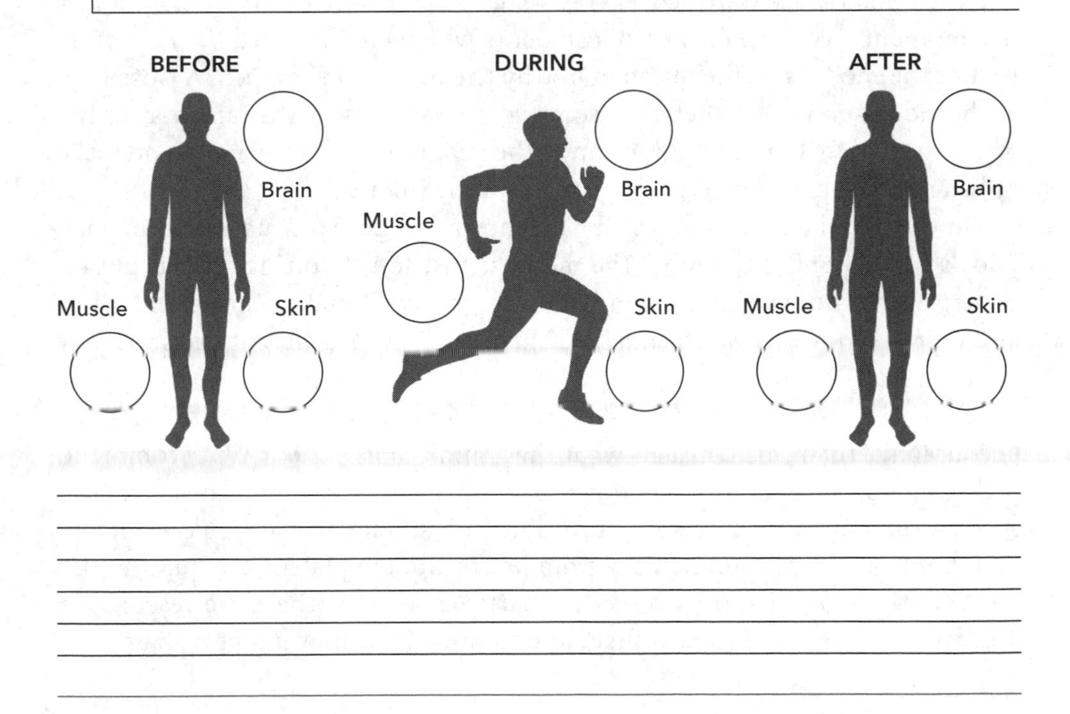

____

____

____

____

____

____

____

in which the students could hypothesize about the runner's core temperature over the course of the race.) These requirements prompted arguments among her students about whether the runner's body was responding to stimuli before the race even started. Her students also puzzled about the runner's later recovery and whether homeostasis simply stops when the body's core temperature, oxygen levels, and blood pH return to normal levels.

In another example, a middle school class was using the imploding tanker car scenario to study the gas laws (recall that the railroad yard workers had steam-cleaned the interior of the car, then mistakenly sealed the top hatch, which led a few minutes later to the tanker's dramatic collapse). The "before" in this case is immediately after the hatch was shut, but prior to the actual implosion. The "during" is halfway through the collapse, and the "after" shows the fully crushed tanker (the model template shown in figure 6.5 has three outlined but otherwise blank railroad cars in these three stages). In our observations of this class at the end of the unit, we saw vigorous debates between groups of students who claimed that "nothing is happening" before the collapse, and peers who countered that there was a reduction in pressure inside the tanker, and the tanker walls were actively resisting the forces that would flatten it in a few moments. We also heard from students who were intrigued by the "after" part of the model and the question of why the implosion stopped. A close look at the video shows that there was some volume left inside the tanker after the event—but why? Can a tanker or any other container be crushed by pressure imbalances so that there is zero space left in the interior?

In the initial template for the tanker shown in figure 6.5, notice the explicit "Goals" statement at the top: "The model is just to get your first ideas out on paper; we are not aiming for 'correct answers.' We'll make our models better and more accurate as we learn more. There are MANY different ways to show your theories."

Each of those three sentences expresses something important to students, encouraging them to represent what they think is going on. We recommend including some version of these goals on all your initial templates. We also include directions that make it clear students should talk with peers before diving into the drawing. At the bottom left of the template is a request to use a convention—that is, to represent air particles as dots (there are reasons we don't use the term *molecules* at first), and to show their movement with arrows.

FIGURE 6.5 **Initial template for tanker implosion**

Name: ______________________________ Period: ______________

Goal for today: With your partner, create an initial model. The model is just to get your first ideas out on paper; we are not aiming for "correct answers." We'll make our models better and more accurate as we learn more. There are MANY different ways to show your theories.

Directions:

1. Talk together and agree on some things to include before anyone starts drawing.
2. In each phase—before, during, after—**draw and label with words** what you **can see** and what you think might be happening that is **unobservable**.

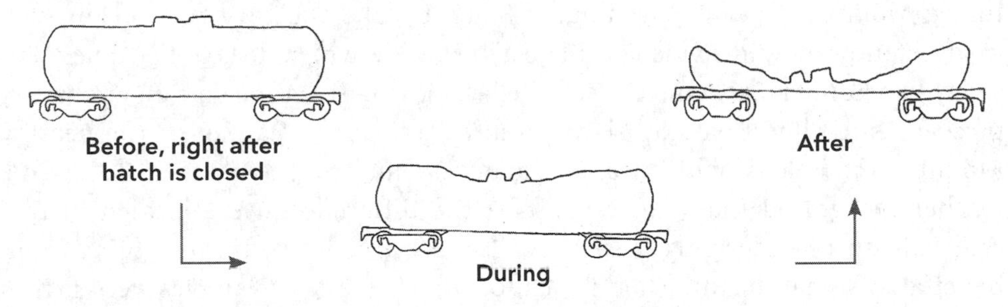

Let's use these conventions:

1. Any kind of air particle should be represented by a dot.
2. The direction of the particle should be shown with an arrow.

Puzzle Box: What questions are puzzling you about this? What would you like to know to improve your model for next time?

Students in nearly every class we've worked with have suggested that the arrow length could then indicate speed. Usually, students negotiate their own conventions when they start revising models later on, but it is important that the class shares a common set, so that they can read and understand each other's representations.

The before-during-after aspect of modeling is challenging for students, but it helps them to make more sense of the phenomenon and to show more of what they know. Early in a unit, the initial three-part models allow the teacher to use students' representations for formative assessment. The models reveal what productive ideas they are using, where their gaps in understanding might be, and what learning experiences should come next.

For some phenomena, we don't use before-during-after. Instead we ask students to compare two situations side by side. For example, one of our seventh-grade teacher colleagues asked his students to model why one member of their class got the flu, but another did not, even though they were twins. Another middle school teacher had planned a unit on plate tectonics and volcanoes, featuring Mount Rainier in Washington State. During the very first lesson, she inadvertently showed a video of Kilauea in Hawaii, which does not erupt explosively like Rainier. Students noticed the lava oozing from the volcano and were puzzled as to why it would behave so differently from Rainier or the nearby Mount Saint Helens, which was decimated by an eruption in the 1980s. The teacher recognized that the reasons for the differences were located in the crustal features underlying these two volcanoes and decided, with urging from her students, that the unit should focus on why these two examples were so different. In these comparative cases, all the same modeling principles apply: the initial drawings, hypothesizing, testing and revising of ideas, and so on. The models, however, are side-by-side analyses.

Make models revisable

Finally, models should be revisable. Because models show how events, things, properties, and ideas are related to one another, students need to test these relationships out. As a result of readings, activities, discussions, and experiments, students make changes to their models over time. The most successful versions of models we've used are drawings on poster paper that can be added to or that can have sticky notes attached as comments (more about this in the next chapter). Remember that students' *ongoing attempts to revise explanations and models are "stretched across" a whole unit* of instruction.

To sum up, models for modeling in classrooms are about phenomena (events or processes) rather than things; are specific to a context (place, time, and situation); are represented pictorially; require both observable and unobservable features; include the before, the during, and the after, or compare cases side by side; and are made to be revisable.

DOING THE WORK WITH STUDENTS: TEMPLATES AND DRAWING CONVENTIONS

The first question you'll want to consider in modeling is "What system of activity do I want my students to represent?" We use the word *system* because models help students think about relationships between parts and actions. If you are having students model a scenario of natural selection, you may want to make sure they represent environmental changes, how random mutations arise and get passed on to progeny, and how variations in traits exist within populations. All these factors work together over long periods of time to cause adaptation and change in the living world. On a more practical level, there has to be room on a sheet of paper for all these features, and space to show how the relationships between them can change over time. If you are modeling a force and motion scenario by showing a car losing traction on a slippery road, you'll have to decide if you want to show both the smaller-scale events happening where the tires make contact with the pavement, and the larger-scale event of the car sliding sideways before coming to a stop in the ditch. Both macro and micro events are important for students to reason about, and it means their models may include both views.

Students also need space to illustrate, label, and explain. We've found that an 11 × 17-inch sheet of paper encourages students to write and draw in more adventuresome ways than do smaller formats. We routinely provide drawing templates for them to use. These might be outlines of the human body if we are studying homeostasis, or silhouettes of batteries if students are studying electrical currents. Templates are standard for our teachers because we don't really want students to worry that they're not artistic enough to render a well-proportioned drawing, or have them spend time on the details of representations that are purely aesthetic. We try to strike a balance, providing just enough structure without constraining students' creativity or funneling their thinking about the science. Even with templates, students find ways to be strikingly original. As teachers, we enjoy whimsical additions to model drawings; we just don't want students' intellectual energies directed too much toward humor at the expense of science.

One exception to the template preference is when you are modeling ecosystems. Because these natural systems are so open to being represented in unique and productive ways, we often just provide students with poster paper and images of resident organisms. We ask students to show more than just "who eats whom" in these models. Their final versions at the end of a unit must show, for example, the movement and transformations of energy, indirect influences that one population has on another, and how the physical environment both shapes and is shaped by the kinds of plants and animals found in the habitat.

Before students begin work on their initial models, you may want to discuss with them some drawing conventions that will be useful. By "conventions," we mean agreed-upon ways of representing things. Commonly shared conventions help students compare their models with others and help you interpret their models more easily. One of the most common conventions used by our teachers is the zoom-in. This is simply a large circle in the model that represents what you would see if you had "microscope eyes." For example, we use zoom-ins to show what is happening in human tissues during cellular respiration, for the interface between car tires and the street to illustrate frictional forces at a small scale, for chemical reactions happening in solutions, and for subduction zones in the earth where oceanic crust dives underneath continental plates. Conventions also include how drawn elements like arrows will be used. Should they represent the magnitude and direction of a force in the sliding car, or should they represent movement and speed? You can't use arrows for both. How will we represent molecules? As particles, or is there some added value to using chemical symbolism (ball-and-stick drawings or molecular formulae)? Having conversations with students about how to represent such things and ideas can, by itself, support their learning of science.

To sum up about conventions, it helps to consider which ones you should discuss with students before modeling (depending upon what is being represented), and which ones should not be specified so that you don't narrow their thinking. These conventions can be negotiated with students at the outset of a modeling experience. We've found it important to have more agreement about these elements as the unit progresses—even to the point of developing a common class key that everyone will use for their models.

Even with all the supports we've described, your students may still be reluctant to commit anything to paper unless they feel it is absolutely correct. This is where you have to counteract your students' deep-seated conceptions of the "game of school." A teacher's framing of this work is very important. Students should be assured that models are *ideas* about how some part of the world works and that their models are *supposed to change over time* as they learn more. It is important that students are able to show what they think in ways that *are different from their classmates*, and that these ideas can later shift or be combined with other ideas. Initial models are starting places that help students see what they might need to learn more about. These are all *stances about modeling and about knowledge production* that should be part of how you frame class activity and should be revisited throughout the year.

Finally, in the interest of equity, students working in groups should hear from everyone before embarking on the drawing phase of modeling. It should be standard practice for partners to talk, with pencils and pens on the table, before sketching a draft, and to all weigh in on the rough-draft sketch before committing to a model. To this end, we've given members of groups different-colored markers and requested that we see everyone's color on the model itself.

HOW TO GET STARTED

You can lower the risks of your first attempts at this science practice by having students model in the context of a single lab activity rather than over the course of a unit. In an earth science class, for example, you might study convection currents that simulate what happens in the mantle to drive plate movement. These currents can be represented when food dye samples of various temperatures are placed in a water bath and observed as they rise and fall. You can provide a simple before-during-after template of this event with just a few prompts, including a request to show what is observable and what is unobservable. Students can do this initial activity without much prior instruction. Then, after studying the underlying science ideas for a couple of days, students can do a second version of their models. In the high school life sciences, students might start with modeling bacterial resistance by culturing successive generations in petri dishes that have a bit of disinfectant placed on half the surface. In chemistry, students might model how first-aid cold packs work in order to learn about endothermic reactions. In any of these cases, students can get used to modeling and you can get used to supporting their reasoning.

As an example of this approach, in figure 6.6, we show a model from a fifth-grade student who had been experimenting with how magnets can repel as well as attract one another. She began to stack them on a wooden dowel and noticed that, even though they were all flipped so as to "push away" from one another, the space between the magnets was not uniform from top to bottom. She and her partner had earlier modeled a simpler

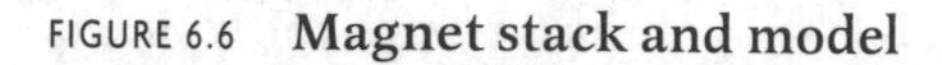

FIGURE 6.6 **Magnet stack and model**

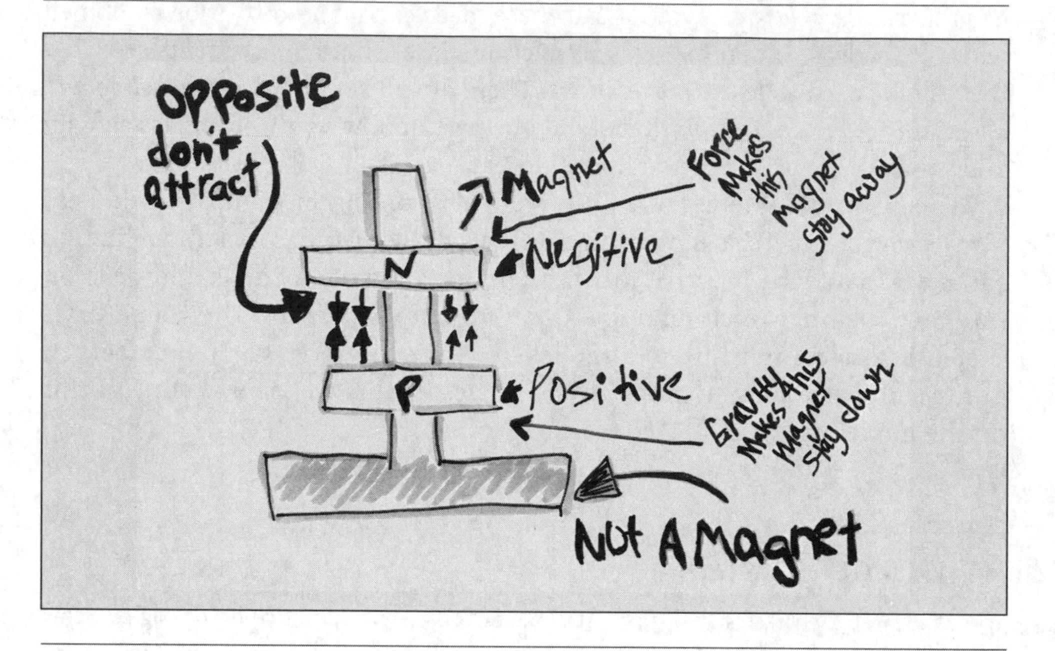

arrangement with just two magnets, but were now trying to understand why, in this new setup, the discs at the top had more room between them than the discs near the bottom. As you might notice, the explanation for this phenomenon is quite complex, involving gravitational forces pulling in one direction, and multiple sets of magnetic fields pushing in two directions at once.

The reason we suggest modeling mini-phenomena is that, under these circumstances, there are fewer moving parts to reason about or draw relationships between. This also means that your template can be simpler. All these situations are more straightforward to model than complex and highly contextualized events, which will come later.

You can use these modeling experiences to try out framing to students what this science practice is all about, how you make different ideas in these initial models public, and how you coordinate learning activities with the science ideas that your students need in order to make changes between their first and second model versions. You can also gain experience looking across many of your students' initial models and seeing what they know, what they need to work on, and, perhaps most importantly, what your template and directions have allowed them to show. These small cycles of modeling may take three or four days, but they can set the stage for more long-term modeling work as you and your students become used to drawing and revising.

SEVEN

MODELING, PART 2

Allowing Students to Show What They Know

IMAGINE YOUR STUDENTS are working in pairs, just finishing up their initial models at the start of a unit. As you circulate around the room, you can see how different sets of partners have approached the task, each emphasizing unique aspects of the phenomenon to draw out and adding unexpected details. Some of these details are just artistic flourishes, but others represent insightful mini-theories that could provoke fruitful discussion if brought out to the rest of the class. The big question now is: *What do I do next?* This chapter lays out a set of plans for "working on and with" these models, just after they've been authored and over the course of a unit. We address the following:

- selecting and sharing out students' initial models
- helping students revise models in response to new evidence and ideas
- ensuring that students can show more of what they know in their final models

SELECTING AND SHARING OUT INITIAL MODELS

Students' initial models can become public resources for reasoning by everyone in the classroom. These resources are not in the form of polished and packaged answers; rather, the models embody creative ideas, in rough-draft form, that reflect the sense-making efforts of young minds and vary in their consistency with known science. They are always worth sharing publicly because it helps your whole class explore questions like: What *could* be included in our explanations? Are there *different or better ways* of showing similar ideas? Do other students have ideas that should make us *rethink* ours? To be effective, however, the sharing of models has to be more than show and tell. We and others have used a strategy for selecting which students' models may be important to share with the class, deciding the order in which models will be presented, and supporting participation by everyone during the actual share-out.[1]

Selection begins early, when you interact with students during their modeling time. As you circulate, you can ask students why they are representing ideas in particular ways or have them narrate what is happening in different parts of their model. But in addition to these exchanges, you can begin assessing each model's potential for stimulating whole-class conversations. You can note the range of science ideas represented in each model, attend to how model elements are drawn, and, in the midst of this quick analysis, consider which students might benefit socially from attention to their ideas. Don't hunt for "correct" or "the best" representations. Often, drawings that are unorthodox, incomplete, or unaligned with canonical science can support productive reasoning by peers more than ones that appear tidy and meticulously labeled with scientific-sounding words.

When you identify a group whose model could stimulate conversation for the rest of the class, you can prepare them by saying, "I like your thinking about [this part of your model]. Would you be willing to talk about it to everyone in a few minutes?" If students are willing, you can give them further guidance; it might sound something like this:

> Okay, plan to take about three minutes; we'll put your model under the doc camera. You can start by sharing how you made the decision to [represent this idea in novel way / include these details / emphasize this part of the phenomenon]. I heard you talking about [this aspect of the model]; maybe that is a good place to start? It would be great if each of you could do some of the talking, and, if you are comfortable taking questions from others, you can do that too. I'll leave that up to you.

After you've prepped three or four student groups, you'll have to consider in what order they should present. Your goal in sequencing is to provoke comparative or imaginative thinking by looking across models, and in particular, by looking at how the same features in different models are represented. To start with, you may want to show a model that displays a few basic features or relationships of the phenomenon, in a way that is easy for others to interpret. This helps the class get oriented to a relatively uncontroversial idea or two that are important to notice as a group. This may be an opportunity to acknowledge students who, despite having created very sparse models, have also captured discussion-worthy hypotheses.

In a fourth-grade unit, for example, one of our teachers used the phenomenon of a flashlight being left on for a few days and gradually "dying." The first student model shared with the class simply showed energy (as dots) located in the battery. This was enough spark for the class to conjecture about what kinds of energy were in the battery and what happened to it—students engaged in back-and-forth reasoning and got the share-out routine off to a good start. This conversation was supported by the teacher, who was prepared to focus students' attention on certain parts of the model, ask good questions, and include students' puzzlements in the dialogue.

Later in your share-out sequence, you might introduce a model that has emphasized some part of the phenomenon that no one else in the room had considered. In this way, you can seed novel and potentially productive ideas throughout the room. In the fourth-grade classroom, one group had prominently illustrated a filament in the flashlight's bulb. This part of the system quickly became a source of interest and possibility for other students in the classroom. What is this thing? What's it made of? What happens to the energy there?

Perhaps the most influential strategy is to select models that represent contrasting theories. In our flashlight example, the teacher noticed one group of students had drawn electrical current as coming into the bulb from *both* directions (figure 7.1). But there was more to their reasoning; they believed that electrons "rub together making their selves hot," that they then "burst into nothing," and that this process continues until the light bulb goes out. Sounds reasonable, and it would indeed explain why batteries use up their energy. This converging pathway was different from every other model that showed a one-way circulating path. The teacher was careful not to judge these drawings, but to highlight the differences and ask students how they might test each model, using batteries and bulbs. It's worth repeating that if you can identify

FIGURE 7.1 **Student's model and explanation of current coming from both directions into bulb (top left); class decision to investigate the "one-way vs. two-way" hypotheses (below)**

Student explanation that accompanied model on left: "… so the electrons go into that support wire and then all the electrons go up each side then when they get to the fillament (sic) the electrons in front go into the fillament and rub together making their selves so hot they make a spark, and where do they go?

After they rub together they burst into nothing then the next go and the next and the next until all of the electrons burst into nothing then that is when the light bulb goes out and you either recharge the battery or the battery isn't good anymore."

"One Way" ← like a race track
Hypothesis

- current leaves one terminal of the battery goes up one support wire, across the filament, down the other side, out to the wire and finally back into the D-Cell

"Two Way"
Hypothesis

- Current flows up both Support wires and meet in filament where electrons rub and get hot, make a spark, glow, + burst into nothing
- continuous series of sparks go really fast so it looks like it's glowing

contrasting but plausible theories in these models, you can set up authentic debates in which students learn to seek evidence in purposeful ways or try out argumentative logic to explain why one alternative seems more reasonable than the other.

Of course, it is never a good idea to set up any group's model as too simplistic, or to contrast it with another model to show that it is lacking or misrepresenting parts of the phenomenon. This could put a chill on all future sharing of ideas. Even in a well-sequenced share-out, we have to monitor our own talk so we don't suggest that someone's initial model is faulty or incomplete, and we steer away from implicit negative comparisons across models.

During the sharing out, you will need to attend to the academic language challenges that your students might face. You may be assuming that everyone understands functional words or phrases like *compare*, *contrast*, *hypothesize*, or *cause and effect*. But inexperience with this language can make even eager students go silent. To build familiarity with such terms, highlight one or two of them before a share-out, perhaps coming up with a definition and example of each. Your class can have a conversation about what it means, for example, to "summarize in your own words" what a model is saying. As you moderate the sharing out, you can point out (or have students point out) when these highlighted terms could be applied.

If you want students to use science language in new ways, you don't want to be the only one talking during the share-out. The audience members should be primed to ask questions of those presenting. You can develop a set of questions that an audience will select from, such as "What does this part of your model represent in the real world?" or "What's happening in this part of your model?" As the student presenters respond, keep in mind that these are initial models, and that students may not have firm convictions about why they've included features or relationships. Conversations at this point in the unit should be framed as "Hey, we're speculating together" and "We're being inclusive about ideas." You can provide time after the share-out for student groups to record two or three ideas from their peers that they feel are worth considering for their own models. As you return to this public sharing routine in subsequent units, your students should be gradually taking over the role of generating and asking questions about models or ideas that they believe are important.

This process of selecting and sharing does not have to be used in every unit. Alternatively, you can organize gallery walks in which all the models are put up in the hallway and students visit them, one by one. It helps to structure this activity so that students aren't simply strolling along and socializing. You can give them questions they can write responses to as they examine the work of two or three sets of peers—for example, "What ideas/features have been included in some of these models that were not in yours?" or "Describe one model that has a different explanation from the one in your model; how is it different?" These responses can later be shared in groups or with the whole class. This routine, as with the select and share-out process, should end with you asking, "What questions do we have now, and what information do we need now to make progress on our models?"

REVISING MODELS IN RESPONSE TO NEW EVIDENCE AND IDEAS

Models are meant to be changed. After a few lessons, students will want to revise some of their original ideas and add new ones. They may want to change cause-and-effect relationships, or question hypotheses that they embedded in their initial models. But at the same time, students are often reluctant to reconstruct models that they've invested time and energy in. This hesitation can reflect beliefs, from years of schooling, that any product created for a class is final, and that if you change your thinking, it means that your original ideas were wrong. Students also dislike the sloppiness of writing on top of their original models, or worse, starting over with a blank piece of paper. There are routines, however, that help make the revision process engaging for students rather than frustrating, and help them see the value of rethinking their models.

We have found that having students apply color-coded sticky notes directly to their initial models is a helpful way for them to indicate how ideas can change with new information, evidence, or logic. Each color (e.g., orange, blue, yellow, green) represents a specific type of comment. As shown in figure 7.2, we use four scaffolds for revision that reflect the ways scientists change models: adding an idea, revising an idea, removing an idea, and posing a new question (notice there are no categories for commenting on how artistic the drawing is or how legible the handwriting is). The students decide what kind of changes they want to suggest, select the appropriately colored note, compose the comment, and then apply it to the model.

With a little experience, students also become capable of offering productive forms of commentary to peers. Early in the school year, we have had them

FIGURE 7.2 **Scaffolds for revising initial models**

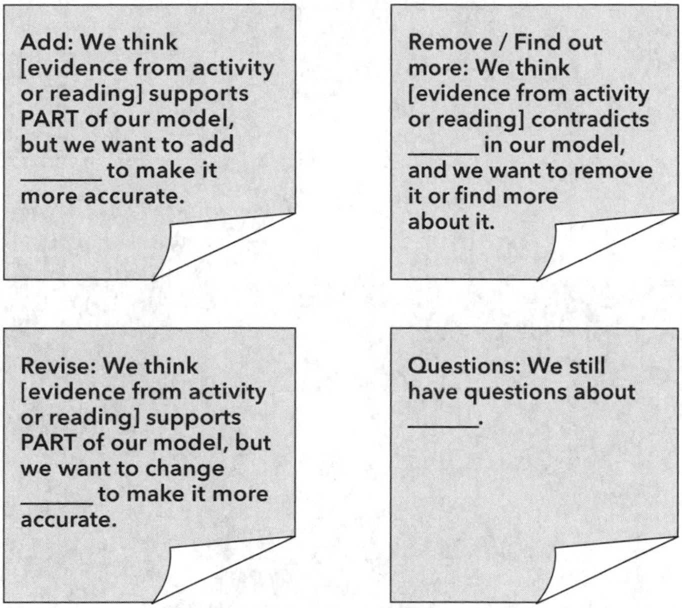

Add: We think [evidence from activity or reading] supports PART of our model, but we want to add _______ to make it more accurate.

Remove / Find out more: We think [evidence from activity or reading] contradicts ______ in our model, and we want to remove it or find more about it.

Revise: We think [evidence from activity or reading] supports PART of our model, but we want to change ______ to make it more accurate.

Questions: We still have questions about ______.

practice by placing notes on their *own* models after a few lessons. They learn how to look at their models, and how to write comments in full sentences that provide reasons for requesting changes. The "full sentences" requirement helps students begin to shape parts of a final written explanation that will eventually accompany their drawn models.

Consider the following example of a model revision. In a high school physics class that we observed, students were studying the video of the previously mentioned urban gymnast who ran up to the side of a building, planted his foot on the wall, launched into a backflip, and then landed neatly on his feet. Students sketched their initial models, and then, several days into the unit after they had completed a number of force and motion activities, they were asked to add commentary to their drawings. The teacher provided the scaffolds shown in figure 7.2.

One group of students had drawn out this event in a creative "time-lapse" style (figure 7.3). Later, they added no less than seven sticky notes to this

FIGURE 7.3 **Original backflip model. Seven revisions added later.**

original model. For example, they included a new idea that came from an activity they had just done: "We think according to our sneakers and socks activity it supports part of our model, but we would like to change that *sneakers matter*, and *traction with the wall and ground matters* should be added to make it even more accurate." This was referring to a race down the hallway, staged by their teacher, between the school's assistant principal and one of the students. The forty-yard dash was unusual in that the administrator was allowed to wear his jogging shoes and the student had to run in his stocking feet. At the start of the race the student, predictably, was slipping and sliding as the assistant principal sped to victory.

After this demonstration, most class members now considered static friction to be a necessary part of the full explanation for the urban gymnast, and some coupled the hallway race with another activity to come to this conclusion: "We think according to the *station activities with the different surfaces*, the type of surface matters because friction matters; the type of surface you kick off of (wall) determines how hard or easy it is to overcome static friction." We can infer from these comments that students saw their lab activities as purposeful because they provided insights about the physics underlying the backflip and introduced science language they perceived as useful (static and dynamic friction, in this case).

These notes also serve to make students' thinking visible to others. As a teacher making the rounds, you may end up visiting the group who wrote the preceding comments and be drawn to their note. You can easily probe their thinking and get them to reason further, working on their ideas by asking, "What did you mean by 'overcoming' static friction? Is static friction good for the jumper or not, and why?" Students may then unpack more of their logic and conclude that overcoming friction might mean that the foot can slide on the ground or on the wall, and that this is *not* a good thing for the jumper.

Using sentence frames on these sticky-note prompts encourages students to think about the different ways that models can be revised, and helps them use scientific ways of talking/writing to express their rationale for possible changes. It also keeps comments from being trivial in nature. In many classrooms we have seen that students not only use these sentence frames but, after a few weeks, even begin to adopt the grammar of these scaffolds in their conversations with peers and with the teacher. The sentence frames are also a way for students to start talking about evidence, and how it could be applied to an explanatory model. To support this thinking, we usually refer to their models as containing a number of claims (like a possible relationship between two elements in a system), any of which could be challenged or supported by evidence. Despite years of work with these sentence frames, we don't believe we have landed on the perfect set. As more teachers do this activity with students, we'll likely see versions that perform especially well under particular circumstances or for different grade levels.

How frequently should you revise models? Through much trial and error in classrooms, we recommend that students return to their initial models to add commentary *once* in the middle of a unit. They should then reconstruct their models near the end of the unit. If you bring out their drawings too often, students will get "model fatigue" and be less engaged with the work. Just to be clear, students draw an initial model early in the unit (day 1 or 2), rethink that version in the middle of a unit by applying the sticky notes to it, and then create a final model and explanation near the end of a unit.

HELPING STUDENTS SHOW MORE OF WHAT THEY KNOW: CREATING A TEMPLATE FOR FINAL MODELS AND EXPLANATIONS

One of the themes running throughout this book is that teachers have to provide more opportunities for students to show what they know. This should be your prime consideration when you prepare students to construct their final models and written explanations. The model template you give students is crucial to this effort. We'll share several supportive structures that allow young learners to express more of what they have learned and enhance the quality of their work.

Provide Directions

One rule to keep at the forefront of your thinking is: "You won't get what you don't ask for." This means that students are attempting a complex performance (model + explanation), but they will not be able to read your mind about what's going to count as a legitimate and complete end product. Directions, for example, should never say, "Model and explain phenomenon X." Instead, you should restate the essential question for the unit and direct students to answer it in their explanation. For example, in our sophomore biology class, the directions to students for the models of Yellowstone's ecosystem changes required them to "Show why so few wolves caused such dramatic changes in populations of other organisms and the physical environment." Similarly, if you want students to cite evidence, you have to ask for it: "In two places on your model, state evidence from an activity or reading we did and say why it supports the relationships that you have drawn." In any assessment task, you want to reduce the chances that some students remain unaware of what you're asking for. Not being explicit puts underserved students and English language learners in particular at a disadvantage. Thankfully, it is easy to level the playing field.

Before students start to work, reinforce that their models will be good for showing some ideas and that their written explanations will be good for showing others. Our view is that it's okay for students' models to show things their explanations can't (such as food webs redistributing energy around the Yellowstone ecosystem), and for their explanations to provide detailed information that is harder to show in drawings (such as how the social behavior of some mammals ensures their collective survival, even in times of stress and diminished resources). We give credit to students when they show some feature of the causal story either in the model or in the accompanying explanation; it does not always have to appear in both.

Designate Spaces to Write Explanations

The final model template differs from the initial template in terms of the space allotted for the explanation. Our teachers have found that it helps to provide cues as to how much writing is expected, such as boxes in which the text should be written. Interestingly, we've found that the bigger the box, the more the students will elaborate in their explanations. Other teachers have gone a step further and included lines inside the boxes (as found in ruled notebooks). This supports handwriting legibility and further hints at how much text is expected. Because most good model templates require students to write—in designated places—about the before, during, and after versions of the anchoring event, this helps them focus on smaller, more manageable chunks of explanation in each panel. It also helps them produce more text in total. None of this ensures that you will get compelling explanatory accounts of the phenomenon, but these simple scaffolds help. Students quickly realize that you expect a certain level of detail and elaboration in their explanations.

Negotiate a Gotta Have Checklist

A checklist of required ideas (described in chapter 12) is incorporated into the model template. We even include checkboxes that students mark off when they've represented an idea from the list in their model or explanation. If you choose, you can request that evidence be cited for parts of the explanation or model, but you will want to specify how many instances are required. Indicate a place for the evidence and reasoning to be written out. Students need practice at this specialized form of writing before they have to do it in their final models (see appendix G). This, of course, is another rule of thumb for teaching: students get better at skills they have a chance to rehearse and get feedback on.

Put All the Supports Together

Let's look at how a teacher sets up students to show what they know, using a model template. This initial template comes from a unit on forces and motion in a ninth-grade physical science class. The anchoring event was video showing a young man who, with the help of friends, attached several dozen large helium balloons to a lawn chair that was tethered to the ground. After all the balloons were inflated, he sat in the chair, and the ropes restraining the chair were then released (we do not recommend this). The man had several cameras aboard, including one on his helmet, and as he rose to an altitude of over five thousand feet, he recorded the changing view below him. At a predetermined point, he took out a pellet gun and began to shoot the balloons, one by one. When his weight became too much for the remaining balloons, he started sinking, then tumbling out of his lawn chair. After a couple seconds of free fall, he deployed a parachute and landed safely. The students were excited by the video and eager to start hypothesizing about how the man could ascend, then fall in a controlled way back to earth.

Figure 7.4 shows the first of four panels in which a pair of students recorded their initial ideas. The four panels represented different points in time—on his way up, on his way down after shooting some of the balloons, in free fall, and then descending with the parachute. Small boxes below each panel were included for students to write their thoughts in. There were no predrawn outlines provided to students; instead, they sketched whatever they wanted in each panel, then labeled it. In the first panel, this pair of students wrote: "When the stops are released he is pulled upward because the helium is lighter than the air and it has an upward force. There is enough helium to pull the weight of the chair and his body up."

These students already have a fairly well-formed scientific explanation at the start of the unit. They have a sense of buoyancy and of opposing pushes and pulls, although they do not use those terms. This teacher looked across all the initial models, and was able to make decisions about when to introduce particular science ideas based on the students' current thinking.

After about eight days of instruction, the students created a final model and explanation. What's important for us to focus on is the way the template allowed students to show what they knew. Figure 7.5 shows the directions the teacher provided.

FIGURE 7.4 **Initial model, by students, of man in balloon chair**

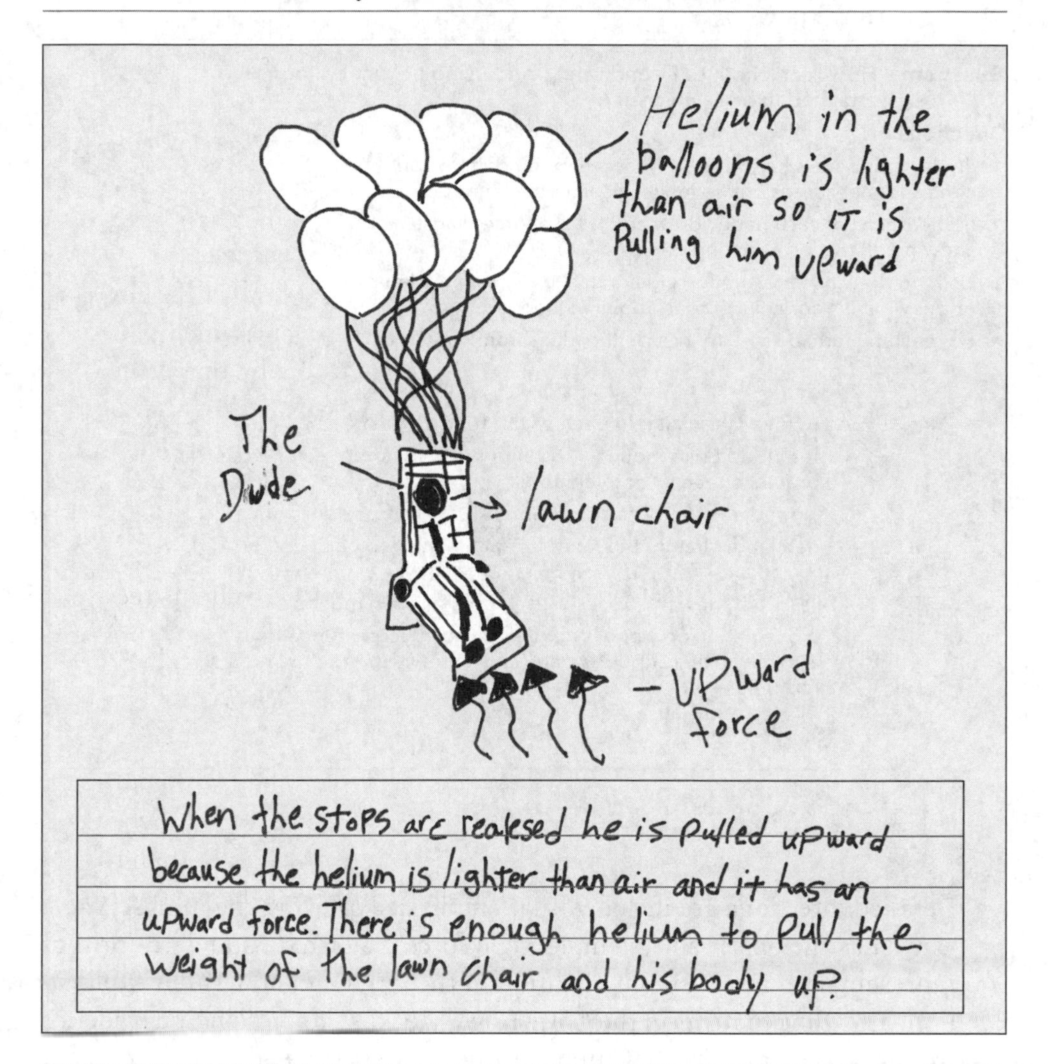

We see several features here:

- There is a clear question that focuses students on what the target of the explanation is. It is not uncommon for students to get wrapped up in modeling and go off on tangents that can be trivial.

FIGURE 7.5 **System of prompts used by teacher to help students "show what they know"**

Question: How is Erik able to fly and safely land using helium balloons attached to his lawn chair?

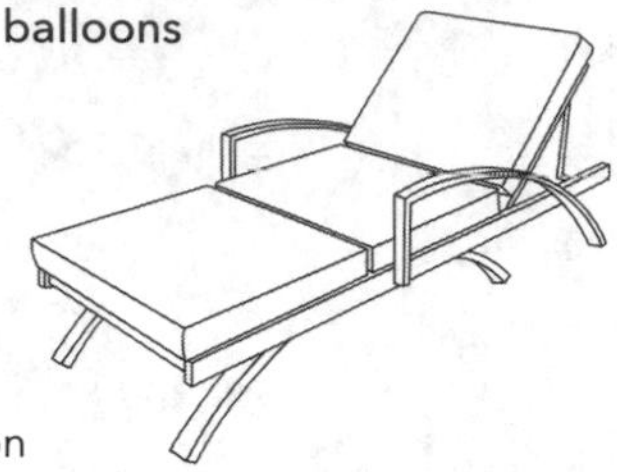

Directions:

1. In the four panels below, draw what is happening that you can't see that causes Erik to move at each point in time.
2. In the boxes of each panel, draw and label a force diagram showing all the forces acting on Erik.
3. Use the drawings and the force diagrams to write an explanation about what is happening at each point in time.
4. For each picture, be sure to include the ideas from the Gotta Have Checklist.

Gotta Have Checklist: In each of the four panels:

- ☐ I described Erik's motion, including direction and type of motion (constant speed or acceleration).
- ☐ I showed all the forces acting on Erik using a force diagram.
- ☐ I identified whether the forces in each panel are balanced or unbalanced.
- ☐ I described how the forces cause any change in motion.
- ☐ On one panel (only one is needed) I provided evidence from a class activity that supports one claim in the drawing or written explanation.

- Detailed directions are included that emphasize drawing the unobservable as well as the observable, completing a force diagram (a standard form of representation in physics) in addition to the drawing and written explanation, attending to items in the Gotta Have Checklist, and providing evidence from one class activity that supports one of the science ideas incorporated into the model.
- The Gotta Have Checklist is prominent, with five items that are more than just requests to use particular vocabulary.

The teacher included a set of mini-directions about how students should provide evidence.

In figure 7.6, the model template itself has more scaffolding:

- The teacher has drawn the outline of the man, lawn chair, and balloons, as well as the parachute in a later panel. This prevented her students from agonizing about how to draw these features and focused them instead on sharing what they knew about the science.
- There is more room (than in the initial model template) to put explanatory text, with large boxes and notebook-style lines to cue students about how much writing is expected.

Importantly, before any drawing or writing happened, this teacher required her students to read these prompts and directions quietly, discuss them in their partner groups, and ask her clarifying questions. It doesn't help to provide all these scaffolds if students don't read and understand them.

Figure 7.6 also shows the final responses (first panel only) of one of the students who drew the initial model we showed earlier. The prompts clearly helped him show what he knows. The student has now written about the "man + lawn chair + ropes + balloons" as a system of forces working together. Nonintuitive science ideas like the normal force are included, as are references to a lab activity that this student feels supported his reasoning about buoyancy. Consider how his thinking has changed since the initial model:

> The forces acting directly on Eric are gravity pulling him to the earth and the normal force of the chair preventing him from falling by pushing up on him. The helium balloons are providing buoyant forces but the balloons are only directly acting on the ropes. The ropes have a tension force that is holding the chair up. Only the chair is acting directly on Eric. In the experiment [we did] with the ball and water, the ball floated on top of the water because it was lighter like the balloon. The ball has a buoyant force.

Are responses like these perfect? No, but most teachers would be excited if their students would write this much accurate explanatory prose for one of the four parts of the causal story, and draw these kinds of models. Note the force diagram on the right side of the panel. These diagrams are the conventional scientific way to express relationships among forces and objects. Students in this class were able to complete these, but the question we should ask is: How little would we know about this individual's thinking if *all he did* was complete the

FIGURE 7.6 **Model several days later, emphasizing man + chair + ropes + balloons as influenced by a system of interacting forces**

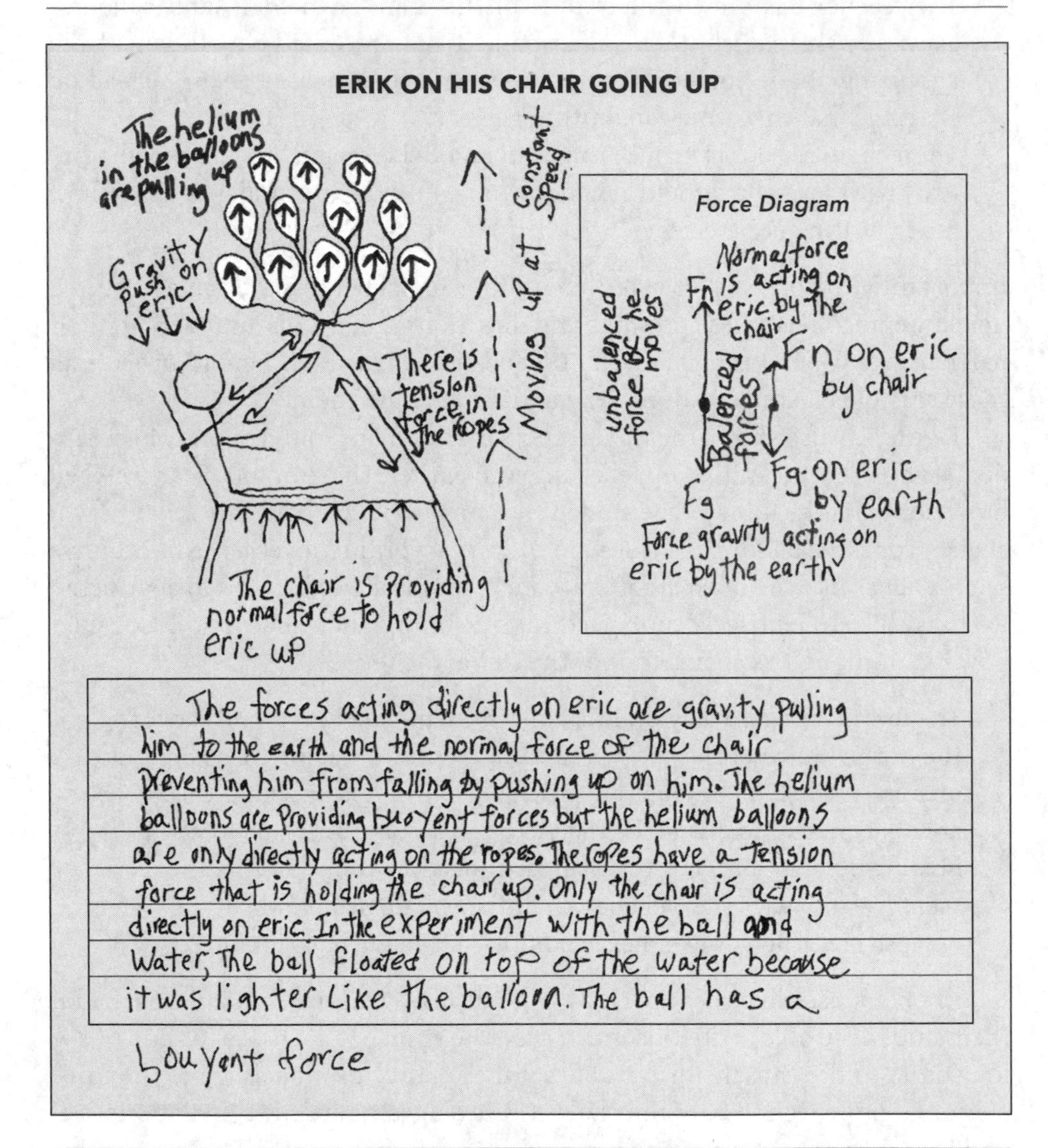

traditional force diagram? The honest answer is that we would see much less of what this student actually understands. If the student drew the force diagram incorrectly, we would not really know what part of his thinking was problematic. With the explanation and model, however, we have quick and detailed insight into how he is reasoning about this event. We should credit the excellent teaching during this unit, but in addition, we cannot overlook the prompts and template that allowed this student to show what he understood at the end.

REMINDERS OF WHY MODELING IS IMPORTANT TO LEARNING

We've explored a lot of specifics in the two chapters on modeling. Let's finish by zooming back out to the bigger ideas about modeling as a scientific practice and how it supports learning. Here are four things to keep in mind:

- *Models motivate students to ask new questions and to propose new hypotheses.* During any AST unit, the experiments that students might suggest, as a result of inspecting gaps or puzzles in their models, have a clear purpose—to improve their own explanations. This helps them decide what kind of evidence to collect and helps them argue for or against different parts of their current models with the evidence they have. These scientific practices (modeling, designing investigations, argument, explanation), which are too often treated as isolated tasks, are now an "ensemble" of meaningful activities that work together.
- *Modeling makes student thinking visible* This is important if one of the big goals of your instruction is to change what and how students think. You can see much of what students bring to the table with their initial models, and this allows you to adapt instruction in a principled way, using your own evidence in the form of student work. You can see what needs to be added to the curriculum, what your young learners already know, and perhaps how to shift the focus of your unit based on students' interests and ideas.
- *Modeling makes students' reasoning available to their peers.* Everyone has access to the ideas of others; these can be borrowed, transformed, discarded, or simply questioned. The process of comparing and critiquing ideas in this way is authentic to the work professionals do, regardless of whether any of these resources become part of a final product. Students also get to hear how their peers talk about features of their models and why they have included

them or taken them out. For example, when they hear classmates reason with analogies, or hear about how evidence supports or refutes ideas, or hear simple if-then logic being applied to problems, it helps them incorporate these rhetorical moves into their own discourse repertoire.

- *The process of modeling helps students see the value of changing their thinking in response to new evidence and ideas.* This stance is not always valued in our current school culture, but Ambitious Science Teaching prioritizes it as a habit of mind. Young learners are enculturated into "improving ideas" as the central ongoing task in their classroom. This improvement mind-set can be cultivated across your school year.

HOW TO GET STARTED

The first time you do model revisions with students, we recommend making it a low-stakes activity for them, without grades or points attached. With this condition in place, it's best to think through some revisions together as a class so that everyone can see and hear the thinking that goes into making changes. Start by soliciting an initial model from a member of the class or a pair of students who are willing to have their representations commented on by peers. Before going public, have a brief consultation with your volunteers about one change they might make to their model, giving them some prompting or support in making this selection. Their choice for change will be the first option you present to the class.

Then, in the whole-class setting, show the model under the document camera and explain the purposes of revision. This is an opportunity to reinforce the expectation that their ideas have to improve over time in response to new information and evidence. This introduction will make more sense if you make available the four ways that a model can be modified (adding an idea, revising an idea, deleting an idea, adding a new question), along with the sentence frames that help students express why a change is warranted. Additionally, it may help if you ask students to recall the activities, readings, and discussions they've had over the past few days; any of these might inform changes to their models.

You can then point out a place on the demonstration model where your volunteer students had suggested a change. If you feel confident that the volunteers can talk out loud about their thinking, let them have the floor. If not, you can take charge. Regardless of who is doing the talking, it is important that the reasons for the change also get unpacked. Perhaps the students had done an investigation, looked at some secondhand data, or had their thinking influenced by what peers had argued about in class. Any of these could be a reason to modify a model, but the reasons for change have to be spelled out. This is one of the ways scientific argument is integrated into the modeling process.

Talk out one example of each of the four types of change and what might be written on the sticky notes. Students will catch on quickly and can join in the reasoning about what to write. You can then get them started on their own models. As you circulate, you should be ready to offer prompts to students who are still struggling with the ideas of changing their models or writing full sentences on the sticky notes.

You must also apply this kind of thoughtfulness when students are first attempting their final models. Because so much of what students end up showing you in their final models depends upon how you structure the template, you'll want to design it for maximum support. Are you giving them plenty of real estate (e.g., 11 × 17-inch paper) to express their thinking? Providing designated places for them to draw and write? Is the phenomenon laid out in "before, during, and after" panels (or, if you have comparative cases, in side-by-side panels)? Did you include the Gotta Have Checklist? Clear directions and a reminder of the question they are trying to address? Requests to use evidence?

The first time students see a template like this, they can get overwhelmed by the very features you intend to support their drawing and writing. We've found it helpful to spend a few minutes at the start of class focusing on what some of these supports mean. We ask students, for example, to restate in their own words what the directions are asking them to do, or what is expected in the Gotta Have Checklist, or what counts as a "use of evidence." Once they get started, your first round of check-ins with them should be simply to determine if they know what the overall task is. Help them make adjustments if they are struggling. If students' efforts are well under way, press them to show you where, in their models or explanations, one of the Gotta Have Checklist ideas is represented. This list actually levels the playing field for students who don't realize what is being asked of them in a complex task—so, for their sake, be a warm demander and point out which ideas on the list they have shown and which ones they haven't.

After the school day is over, invite a colleague to your room and spend some time together with students' models. Spread them out on a table, and take a few minutes just to look. Then apply a more critical eye. Are there science ideas from the unit that show up in most of the models? What errors or omissions show up consistently? If you see gaps when looking across models, you should backtrack to where in the unit you *thought* that you taught those ideas. The teachers we've worked with will often find a problematic area in students' models and then recall that they had done the relevant activities with students but they had *not taken the time* to support their sense making through talk or reflective writing. Sadly, they'll recall: "We did the lab, but we skipped having a conversation about it."

Oops. This is simply another reminder not to assume that activities, by themselves, support learning. Conversely, nearly all our teachers see issues in their students' explanations and models that arise because of the template design, rather than from their instruction during the unit. For example, something you included in the template might be misleading, or you didn't explicitly ask for an idea to be included.

There may be other trends in the models and explanations that pleasantly surprise you. Perhaps students who have struggled in the past to talk or write are now showing more of what they know. This is quite common. Take a moment to celebrate, but also ask yourself: What combination of opportunities and scaffolding allowed them to show so much more? There is a lot to learn from examining final models and explanations, both about student thinking and about how to improve your teaching.

EIGHT

CORE PRACTICE SET #3

Supporting Ongoing Changes in Thinking: Introducing New Ideas

MANY YEARS AGO, we were creating a video documentary of a high school teacher's biology classroom. During the editing phase of the project, we noticed how many times students remarked that their science ideas were in flux; without prompting they would say "He [a peer] changed my mind about that idea" or "I used to think this . . . but now I think this . . ." The regularity of these comments struck us as unusual, because in many classrooms there are subtle messages given to students that you either get it or you don't. In this classroom, however, the teacher was all about changing thinking, and knew it was not an all-or-nothing process. She was intentional about which science ideas were going to be worked on, what her students' opportunities to reason were going to look like, and how she would recognize that their goals were being achieved.

When we refer to supporting changes in thinking, we do not mean providing daily doses of information or alerting students to their misconceptions, then telling them what to believe instead. These methods may sound like the sensible or even responsible thing to do, but the research is clear that the "stamping in" of correct ideas and "stamping out" of inaccurate ones has little impact on learning. Students may be able to repeat key phrases back to you in class or manipulate formulas on tests, but merely adding lots of facts to young minds is like building a house of cards—it seems impressive, but will soon fall apart.

Changing students' thinking requires that they actively engage in sense making. Some of this is done individually and some has to be done with peers. In either case, it involves making connections between ideas, interpreting experiences with materials or data for the purpose of understanding real-world events, and trying out academic language as a way to express what they are learning. Developing new knowledge, however, is not the only goal. We want students to be able to identify their gaps in understanding and the resources or experiences they need to learn more. We want students to select and deploy sense-making strategies for themselves, without depending exclusively on us to direct their every move. We refer to all of this as *learning how to learn*.

The teaching practices in this and the following two chapters (see figure 8.1) provide guidance for helping more students make deeper sense of science concepts, explore these ideas in the context of activities they find meaningful, and apply what they learn to their evolving explanations and models. In the first practice, introducing new ideas, teachers explore with students a science concept or principle that they'll need in order to make sense of an upcoming activity and, more broadly, of phenomena they will encounter or read about. This practice is important because science, as a discipline, has developed a knowledge base that students must use to go beyond their personal experiences and current conceptions of how the natural world operates. We, as educators, are responsible for their understanding of key parts of this knowledge base. In the second practice, students engage in "finding out" activities, where they use these newly introduced ideas to solve problems or conduct investigations. In the process, they make deeper sense of the unit's focal concepts or principles and learn how to participate in different forms of disciplinary work. In the third practice, the whole class acts as a science community to deliberate about "What do we think we know as a result of the activity?" This collective thinking, then, requires everyone to make their reasoning public and open to critique, both of which contribute to learning. The outcomes of most AST activities are not self-evident, in part because they don't funnel students toward right answers. In some cases, they generate more productive questions than they resolve.

The three practices in this set are *used together in cycles, multiple times* throughout a unit as students accumulate knowledge toward goals that are bigger than those of any one activity. To be clear, these strategies are used after you have elicited your students' initial ideas and experiences with the big ideas of the unit, and then used this information to modify upcoming lessons.

FIGURE 8.1 **Core practices: Set 3**

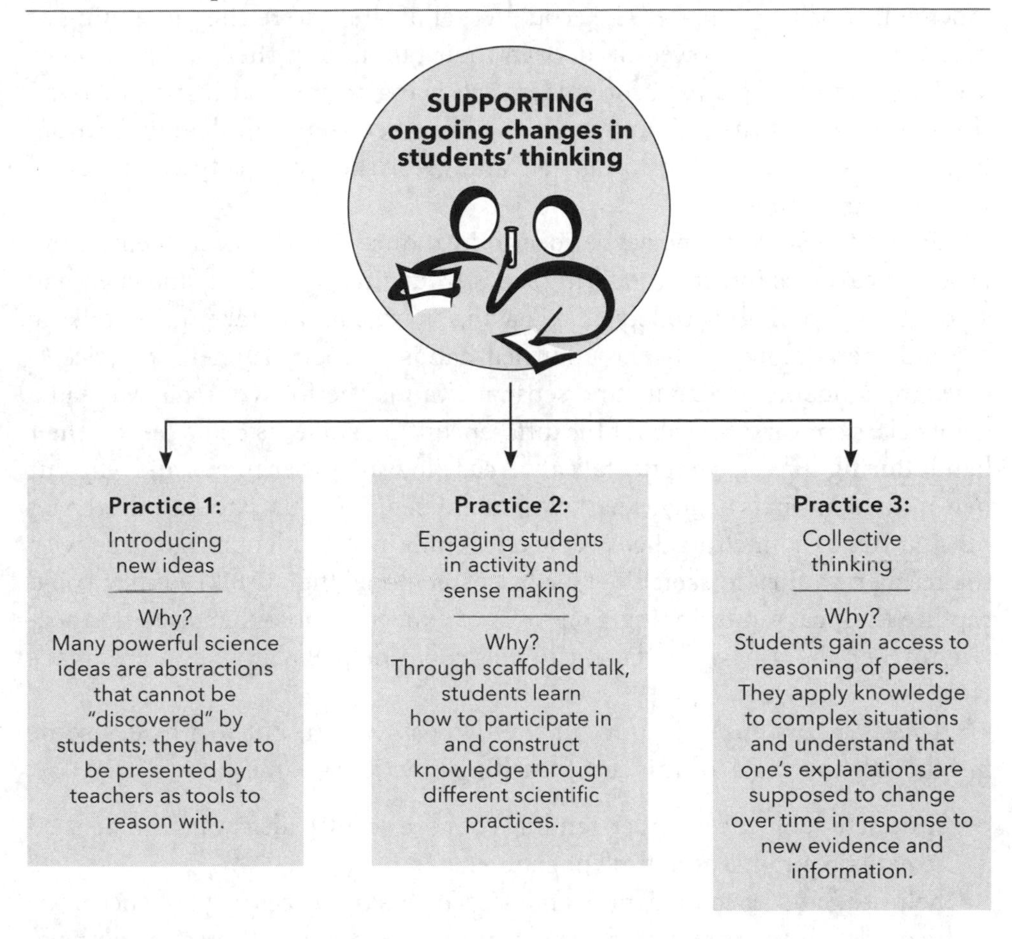

We now provide a quick overview of the three practices working together before diving into each one in more detail. You begin on the second or third day of any unit by selecting a new science idea to introduce. This idea can come from your existing unit plan, or you can reprioritize what ideas to address based on the current puzzlements and partial theories you hear from your students. Here's an example of making minor modifications to the unit plan. In the fifth-grade unit we've described about the singer who could break a glass with the energy from his voice, students began to speculate on the first day about the difference between sound traveling across space and air itself moving from place to place. They couldn't tell if these were the same thing or not. This confusion showed up both in their talk and in their initial models. We had originally planned for students to bring in musical instruments on the second day of the unit and study the links between sound and vibration. Now, however, we wanted to take advantage of their curiosity about sound versus air movement.

So, on the second day, we decided to introduce the idea of air being composed of particles (practice 1 in this set: introducing new ideas) and then had students engage in data collection by having them pair and take turns "talking at" and then "blowing at" the upturned hands of their partners (practice 2: engaging students in activity and sense making). The following day we had a whole-class conversation about the differences that students could feel on their hands during the impromptu "talk at versus blow at" experiments and why our new understandings of air, as particles, might begin to illustrate the differences in wind versus sound (practice 3: collective thinking). Students negotiated with the teacher on how to sketch out a consensus model (figure 8.2) so they could capture their current thinking. Some, as we expected, were then eager to theorize further about how the day's activities could help explain why the singer's voice broke the glass.

This is a condensed account of the two-day lesson, but it provides some insight into the goals for this set of teaching practices—namely, to:

- foster students' deep understandings of core science ideas that can be used to explain a range of natural phenomena;
- help students learn to identify and use various resources (e.g., science practices, data, graphs and other representations of information, one another's ideas, authoritative science ideas) for solving problems and developing new knowledge; and
- support the development of students' academic language, including talk about generating hypotheses, discussing qualities of evidence, comparing and contrasting models, talking about events in terms of causes and effects, and acknowledging uncertainty about knowledge claims.

Our brief example of the three teaching practices comprising this set may appear similar to what you already do; however, when we dive into their details,

FIGURE 8.2 **Consensus model developed by students for "talking at" vs. "blowing at" your partner's hand**

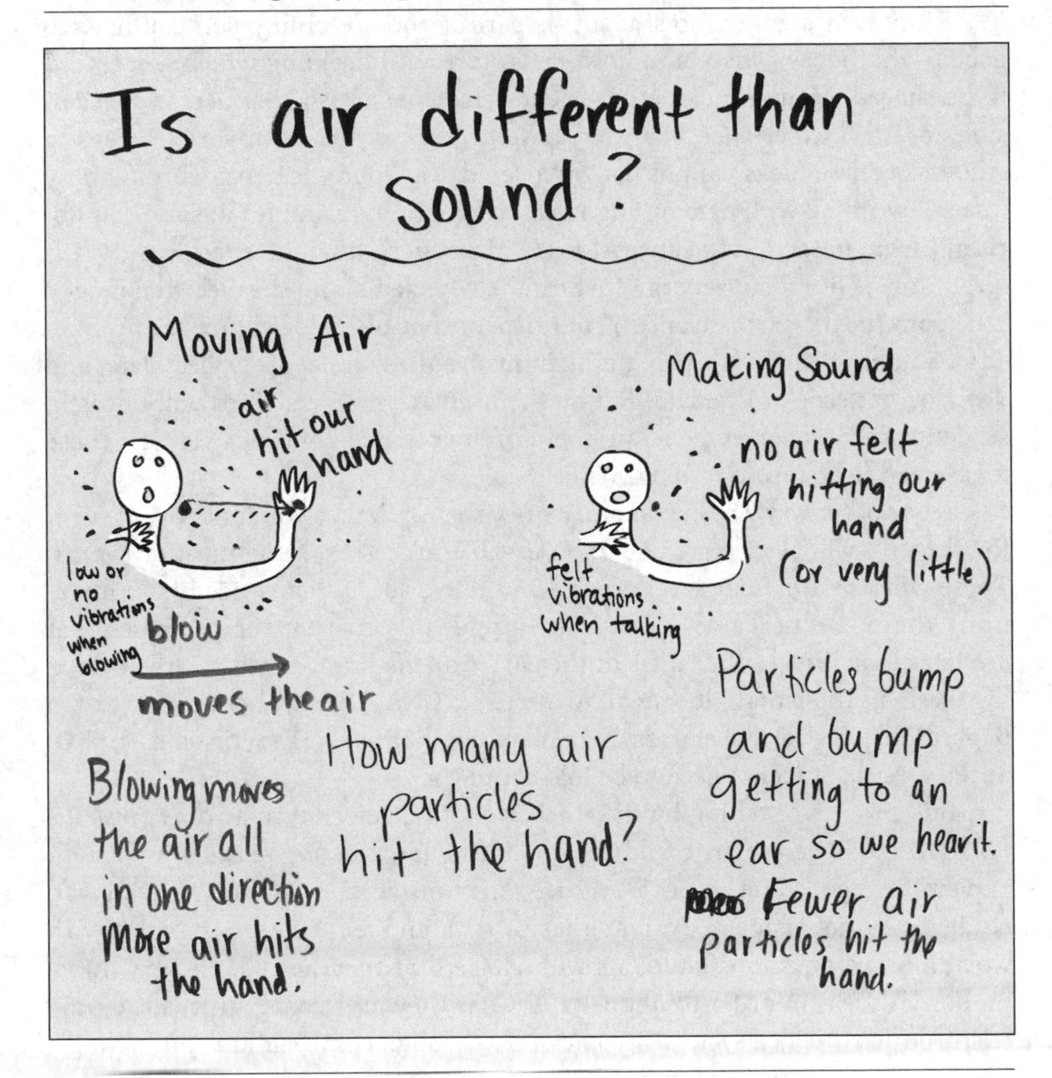

you'll find that they require special skills, tools, and routines that may not be as familiar. This chapter illustrates the first practice (introducing new ideas). Chapters 9 and 10 are devoted to the second and third practices of this set, respectively.

PRACTICE 1: INTRODUCING NEW SCIENCE IDEAS

In AST, presenting ideas to students is part of good teaching; you just have to decide what ideas, when, and how. Let's start with deciding which ideas need to be shared. Many fundamental science concepts are so abstract, so inaccessible to direct experience, that they have to be represented in various ways to students. These ideas cannot be "discovered" by young learners through any form of working with data or unguided observation. Examples include equilibrium (in chemistry), unbalanced forces (in physics), how the sun gives off different forms of radiant energy that can be depicted along the electromagnetic spectrum (in the earth sciences), and the concept of ecological niches (in biology). Students can do lab activities that involve these concepts, collect data, and describe patterns in the data. But by themselves, these activities will not help students spontaneously generate the aforementioned concepts. Rather, these ideas need to be introduced by you.

Put another way, the explanatory story that underlies your anchoring event for the unit will always include unobservable processes, structures, or events. These will explain what *is* observable in the world. Students need help imagining these unobservables because they refer to structures or processes that are inaccessible (e.g., the layers of the earth, or the brain sensing carbon dioxide levels in the blood), too small to see (e.g., DNA, chemical bonding), or too conceptual (e.g., selective pressure in plant populations, magnetic fields, unbalanced forces). Students need to be told about these.

This process of telling by the teacher is often referred to as *direct instruction*, although we prefer to add the qualifier *interactive* for reasons we'll soon explain.[1] A good time to do *interactive direct instruction* is after you've given students a brief exposure to materials or data and let them puzzle a bit with well-designed questions. Students will realize that they need some new information or ideas in order to understand what they are seeing. This is the time to introduce the idea(s), which you can do through combinations of readings, images, video, or teacher presentations. If you choose to do a presentation, plan on allotting no more than fifteen or twenty minutes for it, but this can vary by students' tolerance for listening and by grade level.

Right away a caution is in order: you do *not* want to use interactive direct instruction to create explanations *for* your students. Rather, you want students to know enough about a conceptual idea to use it themselves to talk about and refine explanations of phenomena they are studying. For niches in biology, for example, you can describe what they are, why they are important, and how they work generally in ecosystems. But you should not do all the reasoning for the students and give them exhaustive explanations of how niches function in the ecosystem that the class is currently studying. This is the intellectual work you want students to do.

Following interactive direct instruction, students can "take the idea with them" back into the lab activities. The aim is for students to *use the instructed ideas to reason with* as they do the activity. You can encourage them to integrate new conceptual terms as they explain what they think is happening in the lab or when they start applying ideas learned in the lab to other natural phenomena. You do not want students to engage in activity simply to confirm an idea you have given them instruction on. It can be tricky to strike a balance between structuring the activity so they can do sense making, while at the same time ensuring they can carry out procedures that will produce something to reason about.

Even though we have made recommendations about a sequence for these instructional moves, we want to be clear that the research on learning is *not definitive* about whether the teacher should introduce canonical ideas before any hands-on activity takes place, or the activity should precede the introduction of science ideas. As we will describe in an upcoming section, the order depends upon the science topic and the possibilities of doing unguided sense making with materials. What really matters is how the teacher balances and integrates these two learning opportunities.

Deciding How Much of the Science You Should Present

First, you must decide how much of a science idea to share with students and how much to let them figure out. Here's an example. In upper elementary and middle school units on the phenomena of floating and sinking, students often wonder how something that seems quite heavy (a ship) can float in water, while something that is not so heavy (a ball of clay) will sink. When it comes to these events, the idea of density is helpful, but by itself it won't explain the puzzling outcomes in ways that are sufficiently sophisticated for young learners. At some point, the class must use the idea of buoyant forces to reason about objects that float or sink in water. Buoyant forces, however, are still only part of the explanatory storyline; these forces act in opposition to gravity. Here, then, is the dilemma for the teacher. He is about to do a lab in which students are submerging masses underwater and predicting what they will weigh. The point is to help students understand that there are forces pulling down (gravity), but there are also net forces directed upward on the mass (buoyant forces exerted by the water). The masses thus appear to weigh less underwater.

What ideas do students need in order to start reasoning about floating and sinking? The teacher has several choices, each with a different impact on students' opportunities to reason:

A. He could introduce a full explanation for the phenomenon they are about to collect data on.
B. He could introduce the idea of buoyant forces, meaning that the water exerts forces on objects that are floating or submerged.
C. He could introduce the idea of buoyancy *and* that of unbalanced forces (i.e., when buoyant forces are greater than gravity, an object will float).

The teacher did not choose option A because handing students the full explanation would prevent them from doing any intellectual work—it's all been done for them. The lab exercise would only confirm what the teacher had told them. He did not choose option C because this was also a bit of spoon-feeding. In addition, he thought that, during the activity itself, he could visit student groups and pose questions that would get them to recognize that gravity is an opposing force, and that it is the relationship between gravity and buoyant forces that determines whether objects sink or float. Option C represents an important learning goal, but one that should be reasoned out by students, rather than delivered by teachers and passively absorbed by students. Option B appeared to give students *enough to reason with*—it would give them an idea and a bit of conceptual language to use as they try out possible explanations for the phenomena. The teacher further reasoned that, in an upcoming lesson, students could use the idea of buoyancy together with balanced and unbalanced forces to explain how submarines can be engineered to rise and sink in the ocean. This teacher gave his students "just enough" instruction, in terms of ideas to reason with.

Launching an Episode of Interactive Direct Instruction

Starting off on the right foot is important. Our teacher colleagues usually begin by communicating to students why they are being introduced to an idea. You could use something like the framework shown in figure 8.3. As written, it sounds stiff and formulaic—you wouldn't say these exact words. However, this structure sets up students to understand why they are getting a dose of "telling" and how they might keep a record of the introduced ideas. Every part of this framing serves a purpose.

Presenting Ideas to Students

There are no step-by-step recipes for flawless forms of interactive direct instruction. There are, however, a number of moves that will help students think along with you and resources that make a big difference in how they make meaning of what you're saying. These resources are: representations of the target idea (diagrams, drawings, graphs, computer simulations, flow charts, maps, physical models, or mathematical formulas); analogies or metaphors for the idea; and multiple examples of the idea in different contexts. It helps if you can coordinate the use of *all three* types of resources and make the *connections among them* explicit. Using various representations is particularly beneficial for ELLs

FIGURE 8.3 **Framing for the introduction of a new science idea**

Framing	Purpose
Yesterday we were wrestling with [a puzzle, problem, idea, some new data], *and several of you asked about* [an idea, a relationship, a "what if" scenario].	**Connects to previous day's activity**
These are good questions and you are puzzling about important ideas.	**Positions previous students' ideas and questions as important**
If we don't understand this particular [problem, idea], *we can't make progress in understanding* [a key phenomenon or event that is anchoring your unit of instruction].	**Says how the day's activities can help everyone make progress**
So I'd like to help you now with something called [use the scientific term].	**Introduces key vocabulary in a context**
Here is how I'd like you to use your science notebook in the next few minutes [describe how they can do something other than copy everything you say].	**Provides some structure for how students can record or respond to the newly introduced idea**

in their early stages of language development. These students rely on representations to make sense of complex or abstract ideas, without having to depend exclusively on text or the spoken word; representations enable them to participate in small-group conversations and express their own reasoning to others.[2] Multiple modalities for communicating ideas are not just helpful aids for classroom teaching, they are authentic tools of the discipline for students—scientific literacy is defined, in part, by being able to interpret representations and use them to work on problems with others.

Interaction with students is central to the success of this practice. You should check early whether they have understood the first couple of things you've said; don't assume a lack of questions or puzzled looks means everyone comprehends what you've said. If you don't check, and students don't understand, they will be frustrated when you try to build on those ideas and then really get lost. When they cannot keep track of your explanations or how you are using representations, they'll resort to rote note taking, without doing a lot of thinking. To assist your students in sense making, use "turn and talk to your neighbor," asking pairs to come up with an example of some concept or relationship you've just described, discuss how two parts of an idea might be related, or make a prediction based on the idea. You can encourage relevance by asking students how the instructed idea might apply to situations they are familiar with, or to their lives outside of school. The "turn and talk" lessens the risk of venturing a response to share with the whole class, and when a few students describe what they've talked about with partners, it helps you understand how you might adjust the next few minutes of instruction.

Another means to support active engagement is having students use graphic organizers to structure their note taking. One popular and effective tool is the Frayer diagram, shown in figure 8.4. These tools work best when you are focusing students on a single concept like acceleration in physical science, or cellular respiration in biology. Students can draw this template into their notebooks and fill it as the class conversation progresses. Of course we don't recommend you telling students what to write in every quadrant—they can certainly do the reasoning along with you to fill in the "examples" and "illustrations" containers. We often urge students to write more in each quadrant than is discussed publicly; its purpose is to help them understand the concept using whatever prompts they feel works for them.

FIGURE 8.4 **Frayer graphic organizer for concept building**

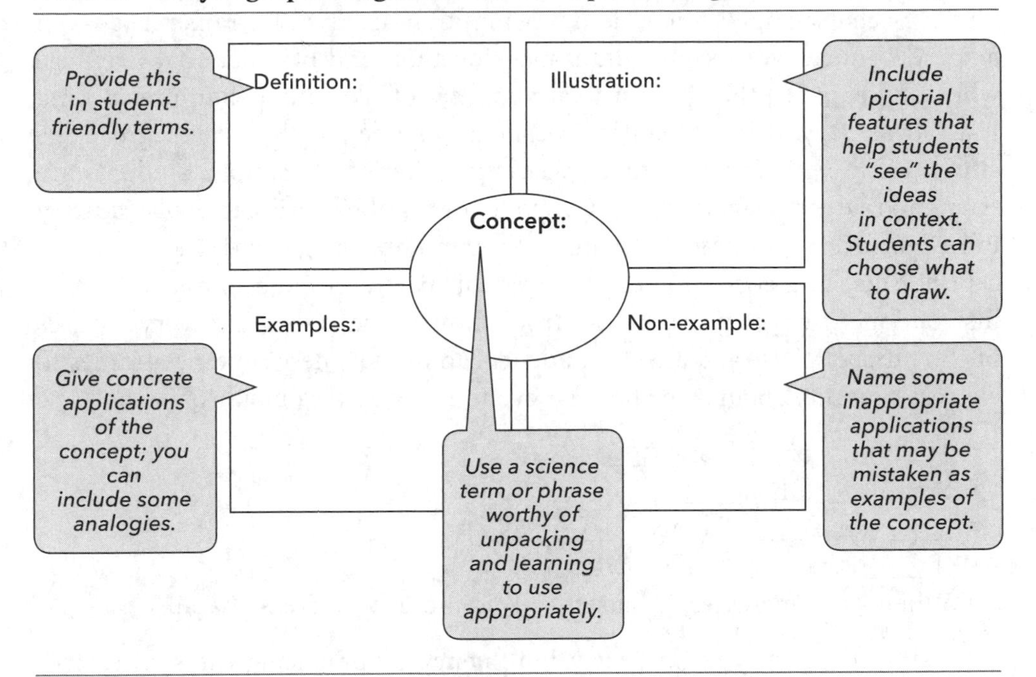

Although filling in this type of graphic organizer and taking careful notes does not necessarily mean that students understand the concept, it does make it more likely that they have a foothold on the idea and can start learning more. Cellular respiration, for example, is not a concept that can be grasped in great depth after a mini-lecture, but students can acquire some beginning language and some concrete examples to carry forward into activities and discussions, where shaky knowledge can be stabilized and linked to a network of other ideas (that's learning).

Not all episodes of interactive direct instruction involve science concepts. Occasionally you will want to help students understand how language is used in the service of carrying out science practices. Here are some examples:

- *Analyze* how evidence supports or contradicts a claim.
- *Describe* how a model shows important parts of a phenomenon or system.
- *Design* an investigation that answers a particular question.

Each of these tasks is an example of *functional language*, meaning communicative acts (saying, writing, doing, being) that are used to transmit ideas in a social context. We recommend unpacking this kind of functional language when students feel they need it to accomplish a task, rather than preteaching it as vocabulary. If it is taught as vocabulary, students will not recognize the situations in which this language is useful. When, for example, students are trying to revise explanations for why populations of bacteria can evolve quickly in response to environmental changes, they might struggle with the basic concept of cause and effect. You can address this partly through interactive direct instruction. After you give a short introduction, perhaps showing everyday uses of the phrase "cause and effect," students can fill out these sentence frames to express causality about some familiar event in their lives outside of school.

- The effects of _____ are _____.
- Since _____, then _____.
- _____ causes _____ because _____.
- If _____, then _____.
- When _____ increases, it causes _____ to increase/decrease because _____.

Using these, they begin to develop fluency by practicing the structure of complex sentences. Of course, we also want to teach students that not all explanations are causal, so knowing when to use causal versus correlational reasoning is important. One of our high school teacher colleagues used the graphic shown in figure 8.5 to help her students link their reading of a text about mutations to the use of cause and effect language. Later in the unit, when students began to use this functional language to understand cancer, they realized that tumors are the result of multiple, interacting influences both inside and outside the body, rather than a simple cause and effect. As one student wrote on an exit ticket (the feedback slips we introduced in chapter 4): "One thing I understand after working today is that the cause and the effect is a whole chain that connects to one another."

Because each episode of interactive direct instruction is so unique, we are not portraying a full example of such dialogue in this chapter. Instead, in table 8.1 we're sharing some helpful dos and don'ts that apply to any presentation of ideas or functional language. Each "do" in this table is helpful for all learners, but it is *particularly beneficial* for English language learners.[3]

FIGURE 8.5 **Graphic organizer for learning functional language in science—in this case, "cause and effect"**

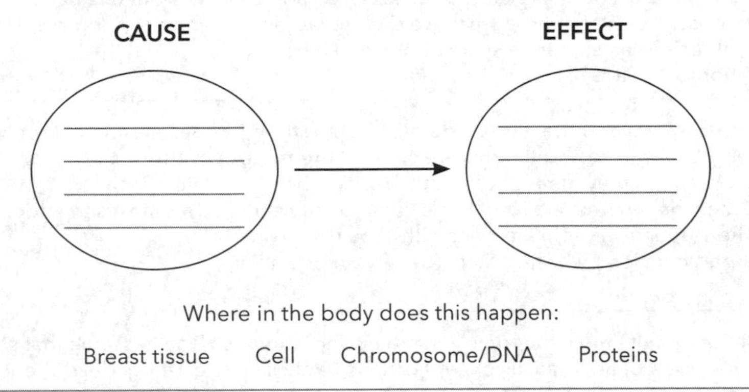

Paying Attention to Your Use of Language

When doing interactive direct instruction, you have to really think about *how* you are communicating, not just *what* you are talking about. The moves we share next are helpful for all learners to understand what you are trying to convey, but again, they are *particularly helpful* for English language learners and special needs students. We don't suggest that these students are any less capable than others; it's just that they cannot use their capabilities when instruction focuses so much on unfamiliar words that come at them too quickly or too ambiguously to process. Try integrating these strategies into your own talk, so that it comes to you naturally.

- Use basic sentence structures that are expressed in active voice (stating the subject, verb, then object of the action), rather than passive voice (object of action, verb, subject). Here's an example of active voice: "The sun [subject] releases [verb] different types of radiant energy [object] from its surface." Here is passive voice: "Different types of radiant energy [object] are released [verb] by the sun [subject]." This sounds like a minor difference, but comprehension issues get magnified as sentences become longer or more complex.

TABLE 8.1 **Do's and don'ts for interactive direct instruction**

	DO	DON'T
Activate prior knowledge	Ask students to recall a previous conversation, activity, reading, or news event that relates to the idea you are about to describe—"Remember when we did the lab on photosynthesis? Can someone share about the kinds of observations we made?"	Don't dive right into new ideas without cueing students' memory or providing a context.
Use vocabulary appropriately	Introduce only two or three new academic terms, rather than many. Describe how the terms are named: Are there prefixes? What do they mean? What is the root of the word? Does the term have a root that is common to other words the students might be familiar with? You can invite student participation by having them identify other similar words.	Be careful not to make your presentation about the memorization of vocabulary. This closes the door on thinking.
Anticipate common alternative conceptions	Consider ahead of time if the topic you are talking about is the likely subject of alternative conceptions. Be explicit and state common misunderstandings to students. Ask questions that might sow the seeds of dissatisfaction about these ideas. This does not work very well to change their ideas (just telling them to think differently), but it is better than not addressing potential confusion.	Don't assume that students lack existing ideas about the topic or don't have preferred ways of talking about these ideas.
Go from concrete to abstract, or familiar to unfamiliar	Start with some tangible thing or event that students have experienced (or can experience together) and use that as the context for talking about the unobservable. This is why a brief round of activity can set up interactive direct instruction nicely.	Don't begin the storyline with abstractions and science vocabulary. You should introduce these later, and only as needed.
Use representations as resources	Use multiple representations of the idea. It helps if one of them is not entirely symbolic or abstract (equations are most symbolic/abstract, for example, and pictures are the least). It's important to help students see how key parts of one representation map to the parallel parts of a second representation. This helps them make sense of both representations and the idea.	Don't use a single representation, especially if that representation is purely abstract/symbolic. Don't, for example, rely on chemical equations alone to talk about the idea of single replacement reactions.
Showing the representations	When you show representations or examples, tell students explicitly what they should pay attention to. For young learners, too many details in some representations (like images, maps, and graphs) can distract and confuse them. If you are using presentation software, you can do a "progressive reveal" that starts with just the basic parts of a representation (like the x- and y-axes of a graph), then overlay parts as students make sense of what is on the screen (like one set of data points at a time).	Don't present or speak about an entire complex representation at once, as if students know what to look for and how to process it.

continued

TABLE 8.1 **Do's and don'ts for interactive direct instruction** *(continued)*

	DO	DON'T
Use analogies as resources	Think of appropriate analogies for the idea you are trying to explain ("air pressure is like a battle between molecules on the outside of a container and the molecules inside" or "homeostasis controls your body temperature like a thermostat controls the temperature in your house"). But also then explain why the analogy maps to the actual event or process you are trying to explain.	Don't use analogies that will lead to misconceptions about the target phenomena. For example, describing heat as a fluid that flows from one place to another actually conflicts with science explanations.
Use multiple examples as resources	Use multiple examples of the idea in action. It helps if these examples are from events the students are familiar with through everyday experience (like breathing hard after a race) or commonly known to students (like how volcanoes periodically erupt). The simpler and more straightforward the example is, the better. Just as with using representations of the idea, avoid distracting details, and help the students know what parts of the example they should to pay attention to.	Don't use examples that are only accessible to students of certain backgrounds, social classes, or ethnic groups. Don't assume all your students have been to the beach or have flown in an airplane.
Get interactive, and check for understanding while you're at it	Midway through the presentation, ask students to restate a new term in their own words, identify what different parts of a representation mean, compare a new idea with one they already know, or come up with a relevant example of the idea. You can also invite questions by offering a sentence frame for them to use: I think I understand ________, but am still confused by ________.	Don't use funneling (fill-in-the-blank) talk as a way to check for understanding. It only tells you if a student or two recognizes the terms you are trolling for.
Summarize at the end	At the end of the presentation, summarize what the new ideas actually help explain. You can do this with students' assistance. Review with students the key relationships, rules of thumb, analogies, models, or metaphors that illuminate the new ideas.	Don't just let the presentation stop abruptly as if students magically made sense of everything.

- Don't use long "wind-ups" to begin your sentences. Here's an example: "Because scientists have quite a few kinds of sensitive instruments that are orbiting the earth on satellites [everything thus far is the wind-up], we know that the sun gives off different types of radiant energy from its surface." Flipping the two halves of this statement would make it easier to understand. Again, it sounds like a trivial correction, but imagine hearing this sentence in a language you are just learning.

- Related to wind-ups are "asides" that are inserted in the middle of sentences. These are super distracting. Here is an example: "The sun, which is just an average size star and about halfway through its life cycle, gives off different types of radiant energy." Can you tell where the verbal bird-walk splits this otherwise sensible sentence in two?
- Don't overuse pronouns—students can't read your mind about what these refer to. Here's an example: "It gives off a lot of radiant energy." If a teacher says this, students will wonder: "What is 'it'?" This sentence was referring to the sun giving off radiant energy, but who would know that? Use pronouns (*it, them, this, they, their, those, she, her, him, his*) only sparingly, or only in cases where the referent—thing or person being referred to—is very clear.
- Use hand and body gestures, but only those that reinforce what you are saying. English language learners pick up on these and use them to make inferences about what is being said. There are thousands of small gestures that actually enhance the communication of ideas. These range from pointing to key parts of representations to using your hands and body to express ideas like "small," "big," "over," and "under." The list is nearly endless.
- If you are a beginning teacher, learn to use your "teacher voice." This means being assertive and making yourself clearly audible throughout the room. Let there be no trace of self-consciousness when you present ideas or interact with students. That's the difference between how a veteran teacher talks and how a beginner talks. While you're at it, don't hide behind the teacher's desk or lab bench to address the class; it sends the signal that you are more comfortable being separated from your students. Get physically closer to the class and diminish that psychological distance.

Some progressive teachers believe that lecturing is to be avoided in the classroom. Other teachers depend too much on the "stand and deliver" method of instruction. Neither of these extremes is appropriate. There are times for telling by the teacher—in specific situations and about ideas that are carefully chosen. In our framework, students will always be hearing about new ideas for the purpose of using them in subsequent activities. In the process of trying out new ideas during activities like conducting investigations, interpreting data, or arguing about evidence, they'll learn more about them—like when to apply them, how to combine them with other ideas and words, and how to use them with peers. For students, new ideas derive most of their meaning from how they help students do particular kinds of work, not so much from their definitions.[3]

HOW TO GET STARTED

Of all the mini-strategies shared in this chapter, we recommend that you try out three or four together. Here's a good set: (1) selecting one or two ideas that you want students to become familiar with and use during an upcoming activity; (2) using two or three different resources (analogies, examples) related to the idea(s) in a way that helps students "look across" them to make sense; (3) inserting a check-in question or task early in the instruction and having a plan ready if students are confused at that point.

It would be helpful if this check-in task or question could give you enough information to diagnose what aspects of the new idea are problematic for your students and why (asking, "Does anyone have any questions so far?" or "Got that?" is not helpful). When you get responses from students, see if you are disadvantaging any particular group. You'll know when certain students stay quiet. Are they all girls? Are they students who struggle with science or are new to the language? Try to figure out why certain groups might find your instruction less helpful. You can also ask students to fill out an exit ticket at the end of class, asking them simply to "Put [the idea] into your own words" or "Write down a new example of [the idea] that we have not talked about in class." It's good to include "What is one question you have about [the idea]?" In these exit-ticket responses, you're looking for quick data trends about your instruction. It is important to identify what you have done that is helpful, and where there's room for improvement. If many of your students are perplexed, look back over table 8.1's dos and don'ts to see if one of those recommendations can help.

NINE

CORE PRACTICE SET #3

Supporting Ongoing Changes in Thinking: Activity and Sense Making

SCIENCE ACTIVITIES ARE a blessing and a curse for learning. On the positive side, students love to work with materials and to socialize with each other in the process. They see abstract ideas come to life through lab work, satisfying the basic human instinct to "try stuff out." On the negative side, some activities can be tedious exercises that amount to little more than filling out work sheets. They may have nothing to do with what students are currently learning, or can be so scripted that no decision making or critical thinking is required.

In this chapter we describe what meaningful activity looks and sounds like, by sharing strategies and tools that deepen students' understanding of science ideas while involving them in authentic disciplinary work. The teaching practice, aptly named, is *engaging students in activity and sense making*. We begin by naming different types of student activity that support learning. Our collection is not exhaustive, but it may help you identify which of these you use often in your own classroom, reflect on what each type of activity is good for, and perhaps expand your repertoire.

In the previous chapter we described how to introduce new science ideas and/or language that would be used by students during activity. But we want to remind you that learning research indicates that, in some cases, it may be best to let students noodle around a bit, puzzle, and explore with materials *before* you formally present new science concepts. Following this brief, perhaps loosely structured encounter with materials or data, students can be introduced to the focal science idea, then use it to view the prior activity through new lenses and employ more precise academic language with peers to make sense of what they're seeing. In this way, students can feel a need for the new idea—the presentation is "just in time" to support their sense making. So, you'll have to use judgment about what comes first: the telling or a brief exposure to the activity.

WHAT COUNTS AS AN ACTIVITY?

We use the word *activity* carefully because the teaching community does not agree on its defining characteristics in the classroom. Activities can include labs, investigations, or just plain hands-on work, and even these mean different things to different teachers. Activities can include everything from doing library research to sorting rocks, from investigating pollution in the creek behind the school building to prepping for the annual science fair. And we have not yet figured out how "engagement with scientific practices" fits in with this assortment. Teachers mix and match these experiences depending on their own backgrounds or on the curriculum they're using.

We cannot resolve these issues in this book, but we can provide names for activity types that place distinctly different demands on students and result in different kinds of learning. At the same time, we make the case that whatever work you ask students to do in groups, you can significantly influence the learning outcomes by attending to a common set of teaching moves.

As you can see in figure 9.1, activities suit different goals. If you want students to develop basic science concepts and abilities, you can have them do proof-of-concept work (#1), develop routine skills by following procedures (#2), use a paper-and-pencil activity (#3), do jigsaw readings (#5), or stage a round of Science Theater (#6). If you want students to engage in authentic scientific work, you might have them design studies (#4) or use simulations to produce analyzable data (#7). If you want them to do authentic work *and* connect science with social issues, you can have students use secondhand data to make and argue claims (#8), or enact a debate that focuses on community issues (health, ethics, economics) related to science (#9).

For activities in any of these categories, you can make *adjustments in the intellectual demand* for students. If you want to provide more support, you can

FIGURE 9.1 **Types of learning activity in science classrooms**

1. Students do "**proof of concept**" work, manipulating materials to confirm a principle or reproduce an expected result.

   Example: Students in chemistry class demonstrate a double replacement reaction by mixing solutions that produce a yellow precipitate of insoluble salts.

2. Students **develop routine skills** by following procedures and practicing.

   Example: Ninth graders learn to balance chemical equations, graph data, use distillation equipment, etc.

3. Students in groups do a **paper-and pencil activity** to make meaning of a concept or system.

   Example: Middle school students use a topographic map to understand what a watershed is and its relationship to distributions of flora and fauna.

4. Students **design their own study** (often within limits set by teacher), collect data, and analyze it.

   Example: Fourth graders use decibel meters to test whether sound moves in all directions from its source.

5. Students work in **jigsaw groups with readings** to identify and share key ideas in texts.

   Example: Sixth graders analyze different web pages for information about forces in sports, sharing the "big ideas" with one another.

6. Students act out "**Science Theater**"–physically representing a science idea with their bodies.

   Example: Seventh-graders take on the roles of blood cells, staging a performance of how blood moves through chambers of an amphibian heart.

7. Students use **computer simulations** to produce observations and data that could not otherwise be collected.

   Example: Physics students create a planetary system to generate data about relationships between mass and gravitational attraction of bodies in space.

8. Students **analyze secondhand data** to identify trends or use as evidence for claims.

   Example: Biology students use data from state health departments to test relationships between reported cases of asthma and family income.

9. Student groups **plan and enact a debate** about science-in-society issues or about what is causing a phenomenon.

   Example: AP environmental studies students participate in structured argument about costs/benefits of converting the country entirely to "clean" energy.

do things like physically model parts of the activity before students try it (how to use lab equipment, for example). You can also model your thinking for students, talking out loud as you search out key ideas from a jigsaw reading and making marginal notes for later reference. Reasoning aloud can also be used to make your thought processes evident to students about designing an investigation, developing a claim, or getting started on any other type of cognitive or procedural task. Support can also take the form of prescribing steps students should follow during an activity, rather than having them make choices of what to do next (this approach can be too prescriptive, so use it with caution). You can provide writing scaffolds, such as sentence frames for parts of the activity, or give students ready-made tools (empty tables for recording observations, or an x-y coordinate plane to plot data points). If students are likely to get lost with a multipart activity, you can require that they ask for a "check-in" conversation with you after completing the early phases and before proceeding further. All of these ways of structuring the student experience allow even the youngest of learners to engage in meaningful and sophisticated work. The supports, of course, are not meant to be permanent—over the school year you would systematically withdraw them and give your students more responsibility to make decisions about how to proceed, create their own tools, and even set their own goals.

There are some types of activity, scaffolded or not, that we should steer clear of. Common "cookbook" investigations, for example, are unhelpful for learning. In these, students mindlessly follow procedures and never get to the sense making required of real science. We also recommend against the investigation of arbitrary questions, even if students are in charge of the design. Science does not involve questions such as "Will my plants grow faster if they are doused with water or with soda?" A question like this, although testable, has little to do with the development of any coherent understanding of underlying causes. In other words, the results do not help develop any serious explanation about plant growth. This applies to first graders as much as to upper-level high school students. Similarly, we advise against investigations outside the bounds of the natural world. The sciences do not investigate questions such as "How many of our students walk to school versus take the bus?" or "Does our cafeteria recycle more than 50 percent of its food waste?" Although these can be motivational hooks for students, they are essentially inquiries without science content.

During activities, it can be tempting to become overly focused on "variables" talk, procedure writing, error analysis, formulaic lab report writing, and the like. These priorities have been an accepted and routine part of school science for decades, but an exclusive focus on them can trivialize what science is. Lab work is most beneficial when it immerses students in authentic disciplinary practices like modeling, designing investigations, analyzing and displaying data, seeking out relevant information for a project, and representing ideas and findings for an audience. Science is more than identifying the variables and writing conclusions.

Regardless of which type of activity is used with students, it should always support some part of an explanation for the unit's anchoring event. Conversely, for every part of that underlying explanation, you should devote at least one learning activity and the sense-making conversations that go with it. Whatever activity you select, it should provide a shared experience for students, around which a common language and set of ideas can be developed.

SENSE MAKING IN ACTIVITY

As much as teachers and students enjoy activities, research on instruction shows that activity, by itself, does not reliably influence learning. What is usually missing is the sense-making talk. Sense making means that students gain insight into some relationship between ideas, representations of those ideas, and experiences they have, but getting this to happen by design in classrooms is difficult. Lots of responses by students can give the mistaken impression that they've made sense of ideas, representations, or explanations; such false indicators can include using technical vocabulary, completing procedural labs, and using formulas in prescribed ways to come up with answers. Sense making is not as simple as this, and it takes time. It is about making connections and recognizing particular kinds of relationships or patterns.

While there are an infinite number of ideas in the world, there is a smaller set of ways that the human mind seeks out links or storylines to make new information meaningful. These more elemental targets of sense making are handy because they can be used in different circumstances to help students learn about a wide range of topics. Some of these targets include *comprehending a conceptual category* and *why something fits in that category* (such as why whales belong to the category of mammals, or why combustion is an example of an exothermic reaction), *recognizing distinctions among things that appear closely related* (e.g., alleles versus genes), and *understanding why things in the same class can be similar or distinct* from one another (e.g., how copper wire, water, and the human body can all be conductors of electricity). In other cases students can *learn how parts of a representation symbolize something in the real world* (for example, how peaks and troughs in a sound wave correspond to places where air molecules are compressed against one another or pulled apart). Students can *recognize how parts make wholes that have unique characteristics* (e.g., cells making up tissues, which make up organs and organ systems, which in turn make up organisms), or *grasp how a new idea can be used to explain an everyday event* (such as buoyant forces helping explain why ice cubes float in water) or *why a set of events or conditions causes something to happen* (e.g., the gravitational tug of the moon causing the earth's tides). These are just a few general categories for how learners process new information and, in some cases, reorganize what they believe.

Sense making is both about understanding an idea (such as mitosis in cells) and using that idea to explain events in the world (why out-of-control mitosis allows some cancers to spread more rapidly than others). We don't want students to come away with only academic understandings of concepts (such as their definitions and examples); we also want students to try to solve authentic problems, *using the ideas as tools*. Even better, we want students to recognize which ideas are relevant as resources when doing intellectual work, without being prompted by the teacher. Great teachers set up conversations that lead students to make these connections for themselves. When students are always told what the relationships are, they are "borrowing" the sense that someone else had made, and consequently their learning is both fragile and temporary.

THE PARTS OF THIS PRACTICE

Sense making for students does not come as a natural by-product of activity; you have to be intentional about integrating opportunities for students to reason about ideas, representations, and experiences throughout the lesson. Here we'll focus on the moves you can make that support sense making by students. We cannot provide detailed recommendations for how to structure or scaffold all the possible types of activity that could be done in a classroom; however, we've selected an example and use it to emphasize strategic moves that make the activity meaningful for students.

Framing the Activity

The way you frame the activity shapes expectations in students for how they'll approach the task. Remember that framing is different from giving clear instructions. We recommend that your framing describe to students how the activity can help them advance their thinking about a particular puzzle or part of the eventual explanation for the unit's anchoring event. Your framing should also include a statement about how they'll use ideas—perhaps they will be making evidence-based claims as part of the activity and need a reminder that such statements use available data but also make inferences about events or relationships that go beyond the data trends. Whatever your activity, provide written and verbal guidance about how to proceed. As we mentioned earlier, you can physically model parts of the activity that could otherwise hang up students and prevent them from getting to work (you can show equipment they'll use or provide examples of completed student work).

Moving Among the Tables: Supporting Insights or Breakthroughs

After framing and directions, your students can begin to work in pairs or groups. This is a time for you to circulate, listen, and then press them to reason further about the science. The purpose of moving among the tables is not to ask students, "Have you got any questions?" or "How far along are you?" Rather, it is to find out what ideas or skills they are wrestling with and to press them to think more deeply using the resources available.[1]

Because you have to be responsive to students' thinking during your brief visits, we advise using *Back-Pocket Questions* (BPQs). These are index cards on which you write different prompts or questions to ask students as you move from table to table. On this card we recommend that you include two questions you'd ask student groups that are having trouble getting started with the task. These might be prompts for them to focus on a specific part of the task first, or to recall earlier class conversations in which a science idea, relevant to the present task, was discussed. Another two questions on the index card could be written for students who are zooming along and need more demanding questions to keep them reasoning together. Then, on the bottom of the index card, you can write generic follow-up questions to students such as "Why do you think that?" or "Do you all agree?" or "Can you unpack that idea for me?"

Having BPQs in hand may sound prescripted and not responsive to students' talk, but just the act of writing them up ahead of time helps you prethink what sticking points your students might face, or develop prompts that encourage groups to think more deeply together. It provides good questions to choose from in case some of your students need higher levels of challenge. Most of our teachers use the questions right off the BPQ card, while others improvise on the spot, using what's on their card just to spark their thinking.

Moving among the tables follows this general pattern: move to a new group of students; listen to their conversation; select a way to focus, redirect, or press on their current thinking through dialogue; make eye contact with every member of the group, asking those who have not contributed if they want to add anything; and then ask students a "leaving question" that keeps them talking after your visit is over.

Advice to novice teachers: you have about three minutes to visit each table group, so don't park yourself at a table for a lengthy conversation. When you enter a group, get down at students' eye level to talk with them rather than loom over them. Don't turn your back on the rest of the class—keep a watchful eye on the room.

The following example of moving among the tables is drawn from our eighth-grade unit on the gas laws, using the imploding railroad tanker car as the anchoring event. Students had just received interactive direct instruction (the first teaching practice in this set) about kinetic molecular motion and phase change. Following this, their teacher provided a demonstration in which he poured an inch of water into an empty soda can, then heated it to a boil on a hot plate. He inverted the can into a water bath, and it immediately collapsed with a bang. The class discussed briefly how the can was like a physical model of the tanker and also the ways that it was unlike the tanker. Then came the challenge of the day's lesson: students were asked to design their own experiment with the soda cans. They had to consider what they learned during the mini-lecture and what they had drawn on their initial models earlier in the unit. The teacher handed these models back to the students, and asked them to keep in mind that their investigations had to test some part of their theory. They would be responsible for articulating this connection to him during the table visits (see appendix F for a written guide that supports sense making during the design of investigations).

Box 9.1 shows the BPQs that the teacher drew up for this lesson. He anticipated that, despite clear instructions, some student groups would be confused about where to start or what was being asked of them. For these encounters he sketched out two questions. Below those he had written other questions for students who were moving forward on the experimental design, but needed to be pressed further about the science. Below that he placed generic follow-up questions that could apply to any group he visited (recall from the discourse chapters in this book that follow-ups are important in verbal exchanges with students). He also reserved space at the bottom of his card for names of students he would ask to share their ideas with the rest of the class later in the lesson, and a reminder about how to "prime" them to talk to their peers. We'll return to the idea of priming, which sets up whole-class discussion, in the last practice of this set.

BOX 9.1 Back-Pocket Questions for table talk about the tanker

Helping students get started

- What do we think was inside the soda can before we turned it upside down in the water? Why?
- Let's start by talking about what we think may be happening inside the soda can we used for the demonstration about the science.

Pressing further

- What will your experiment tell us about the tanker?
- When you say "pressure," what do you mean?

Follow-ups

- Can you say more?
- Do you all agree? Why?
- What makes you think that?

Who will share out about their experiment and rationale?

During small-group work, the teacher prompted his students to use the ideas and the language of kinetic molecular motion in conversations about their proposed experimental designs. How students used these concepts was up to them, but they had to justify their inquiries using facts and principles. The following transcript represents a visit to a group of three students who were trying to test the idea that phase change inside the tanker—from vapor to liquid water—caused the implosion. The teacher is trying to get them to reason about why the molecules "lose energy" to start the condensation process. In this transcript, the teacher has just moved to this group's table and takes a seat in an empty chair. In the center of the table is the initial model of the tanker that students had drawn the previous week.

1 *KATIE:* We could have one soda can that we put in the water bath and one
2 that we just let sit there to see if it collapses. Maybe it would still collapse
3 if it just cooled down after a while.
4 *ABANU:* We could measure which one collapses more, or which one
5 collapses first. We could time it. We could see if they crush the same.
6 *KATIE:* Both of them.
7 *TEACHER:* So, which of the two soda cans in your experiment [points to
8 their model on the table] is like your tanker model here?
9 *ABANU:* We saw in the video that it was raining when the tanker collapsed
10 [sketches raindrops falling on top of tanker model as he talks], so we are
11 thinking it's like the soda can we dunk in the water—
12 *TEACHER:* [Makes eye contact with Mira and Katie] Do you agree with
13 what Abanu said?
14 *KATIE:* [Pauses] Yes—it's like cooling off really fast in the rain, and it makes
15 the steam condense so it takes up less room.
16 *MIRA:* Inside, yeah.
17 *ABANU:* It cools off fast, it makes the steam condense, and then the
18 pressure goes down inside the tanker, or ummm, in the pop can.
19 *TEACHER:* Mm-hmm. Well, I'm kind of interested in the soda can that you
20 heat up and just leave sitting there. Katie, you said it would cool down
21 more slowly and maybe collapse later, but I wanted to hear about how
22 this "cooling off" happens. If we could see with microscope eyes what's
23 going on, what would we see?
24 [Silence for 10 seconds]

25 *TEACHER:* Okay, let's think about the jigsaw pieces you just read, especially
26 the one on kinetic molecular theory. What was the big message from
27 that reading about how molecules move?
28 *KATIE:* That molecules are always in motion?
29 *ABANU:* Hotter molecules move faster—eh, hotter things have faster-
30 moving molecules. So cooling is when they slow down. So phase change,
31 when liquids heat up they—the molecules—move faster until they break
32 away and get into steam.
33 *TEACHER:* So the tanker molecules have to do the opposite of what you just
34 said—they have to slow down. But I don't get how these molecules in the
35 tanker just slow down by themselves; do they just run out of gas?
36 [Laughter]
37 *KATIE:* They lose energy?
38 *TEACHER:* But that doesn't tell me *why* they start slowing down. What
39 makes that happen? Let's look at this part of your model [circles a small
40 part of the tanker wall]; what's going on here?
41 *ABANU:* They're hitting the wall. Maybe that's where they are losing energy?
42 *TEACHER:* Do things lose their energy when they run into walls or the
43 sides of containers?
44 *KATIE:* No? Well, yes, they can. A ball won't bounce back to you all the way
45 if you throw it against the wall. Or like a car crashes and it loses energy,
46 it stops; I suppose it crashes into something. But is that how things cool
47 off? Like at the molecule level?
48 *TEACHER:* Mira, do you want to weigh in on this question of why the
49 molecules lose energy? Cool down?
50 *MIRA:* I agree with Katie.
51 *TEACHER:* About . . . ?
52 *MIRA:* About when things hit a container they lose energy and that energy
53 goes into the container or into a wall. If molecules hit a wall they can
54 bounce off it, but maybe not as fast.
55 *TEACHER:* That's an interesting idea about what might happen to the
56 energy of a molecule in a container—hitting the wall, I mean. Would
57 you all be willing to share your theory later, when we have a discussion
58 about the results from our experiments? Just say, "Here's what we were
59 thinking . . ." It would help everyone in the class.
60 *STUDENTS TOGETHER:* Okay, sure.

61 *TEACHER:* Okay, so for right now, I want you all to talk about your theory,
62 get clear about how or why these molecules are losing energy. The water
63 molecules are starting to stick to one another to create this condensation
64 inside the tanker; we can all agree on that based on what we know from
65 our last unit on phase change. But you need to be able to say *why* those
66 molecules are slowing down in the first place, and where that energy is
67 going. You already have a good start. [Leaves the table]

What did you notice in this segment of talk? The teacher is asking good questions, yes, but let's push our analysis further. He started by listening, and even in those first brief moments he could tell that these students would benefit from being pressed a bit. He was responsive to the ideas he heard (lines 1-6) from two of his students, using those as the basis for his question about how the proposed experimental setup was like their tanker model (lines 7 and 8). He focused on one important idea—how gases "cool off"—and didn't jump around to other topics, nor did he funnel students into saying any particular words or phrases. Instead, he wanted to hear their reasoning (for example, lines 19-27).

He made clear efforts to involve a student, Mira, by turning his attention (his gaze) toward her (lines 12 and 13) and asking if she agreed with her group, then later reserving space for her to comment on the group's working hypothesis (lines 48-50), even to the point of asking her to elaborate on her initial response (lines 51-54). The teacher asked valuable follow-ups at lines 12-13, lines 38-40, and line 51. Each instance made the student's ideas the object of discussion. He also made an important move at the end of his visit (lines 61-67). He was laying the groundwork for the upcoming whole-class discussion about the design of their experiments. Teachers know that getting students to contribute to these public conversations can be difficult, so we recommend preparing two or three groups to kick-start the talk by getting ready, while they are still in small groups, to share their ideas later. In this case, the teacher gave these students some hints about talking to the whole class so that they could make decisions ahead of time about what to say to their peers and how they would say it. Again, we refer to this move as *priming*.

In this round of visits, the teacher did not focus on the details of experimental design, such as identifying variables or how outcomes could be reliably measured. Rather, the science itself was discussed. We don't suggest that investigational procedures are unimportant, but we also recognize that talk about measurement, error, and the like can be premature when students don't first clarify what they are choosing to study and why.

This exchange with one group of students took about four and a half minutes. In a classroom with, say, nine groups of three students each, you could have these conversations with everyone in a standard class period. You'll be thinking on your feet the whole time, but you are ensuring that all of your students have been intellectually challenged that day.

Here's another set of example BPQs that are responsive to student thinking, from the tenth-grade ecosystem unit on the reintroduction of wolves to Yellowstone National Park. Groups were given graphs of different animal populations and plant growth (figure 9.2),[2] then given these prompts to help them start:

- If you had to describe to a person what the trends were for one of your organisms, how would you communicate that? You could use terms like "increasing over time," "decreasing, then increasing," and "increasing slowly, then increasing more rapidly." These are not the only ways you could describe the trends, but you can use them if you want to. Discuss this with the members of your group.
- Now look at the wolf population graph. Discuss with your group what is represented on the x- and y-axes, and if these are different from the axes on the graph of your organism.
- Refer back to your organisms' fact sheets. Are there any links between the wolf and other organisms? These links might be direct, or indirect, or you may see no links at all. To help you, here are some questions to ask yourselves: *How might changes in the graphs of wolf population and another organism be related to one another? For example, does the wolf population affect another species over time? Does that species affect the wolf population over time?*

The goal here was for students to use the graphical data in figure 9.2, along with other information, to continue to revise and build explanations for how the reintroduction of wolves changed the ecosystem. The teacher recalled how some of her students the previous year struggled to understand what the graphs meant, while others were successful in reading the graphs but overlooked basic information they had been given earlier about the organisms as they tried to theorize about what was happening. Knowing this, she sketched out BPQs that could help groups with either of these issues.

FIGURE 9.2 **Yellowstone population graphs**

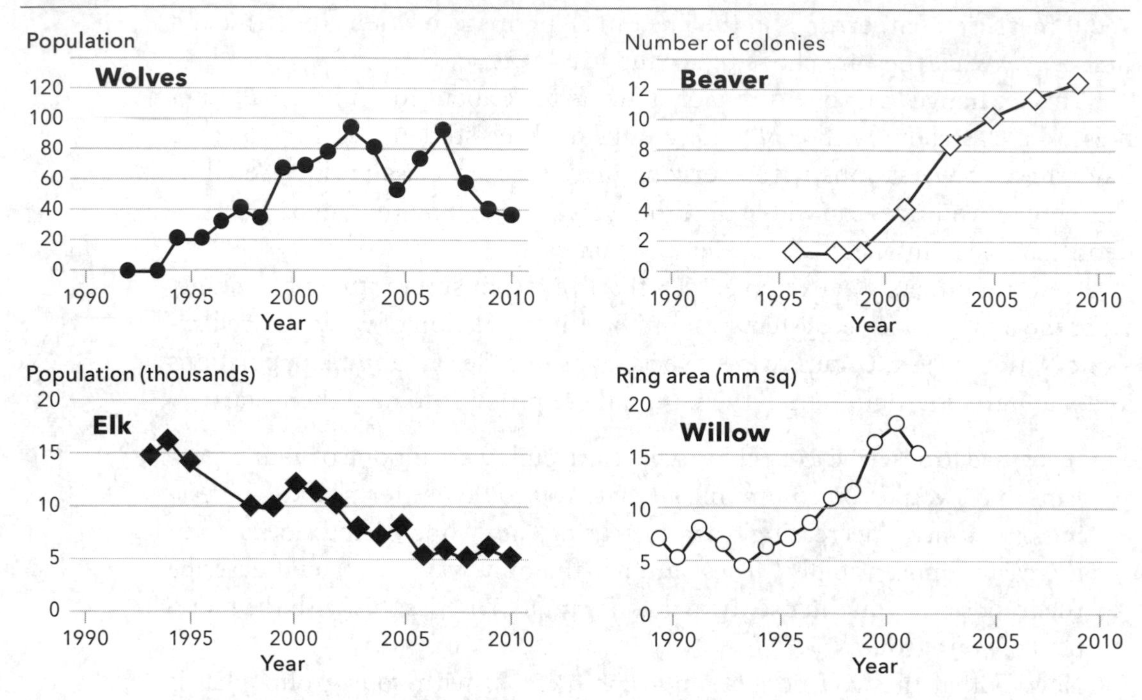

Source: Reprinted from William J. Ripple and Robert L. Beschta, "Tropical Cascades in Yellowstone: The First 15 Years After Wolf Reintroduction," Biological Conservation 145, no. 1 (2012): 205–213, with permission from Elsevier.

As students began to work, she approached one table and listened for a couple of minutes. The students were becoming frustrated by not understanding the graph of willows, which measured not population, as in the other graphs, but the average diameter of their branches (elk browse on willows, not killing them but preventing their growth). The teacher sat down with her students, clarified what the y-axis represented, and then said, "Let's start by looking here [placing the tip of her pen on the first data point in the series]. If we look across at the y-axis, and then down at the x-axis, then what is this one point of data telling us? Just this one?" Students were able to agree that "at Year Zero, the average willow branch researchers measured was about seven square millimeters in area." A second question asked students to look at a later data point in the series and describe what it represented. The students remarked, "Ten years later the average branch was about eleven square millimeters in area—okay, okay, we got it." The teacher, however, persisted: "What does it mean that these numbers are going up? Don't try to tell the story about another organism here; just describe what's happening to the willows if these branch measurements are going up."

She left them with that question and moved to another table where a second group of students had cut out their graphs and aligned them vertically on the table to see more precisely when certain trends began to show up for each organism. This was a good sense-making move on their part, but after a few minutes they were stalling out on how to create a story about the ways in which these populations were influencing each other. The teacher listened for a couple of minutes, then offered, "So, you are saying that the wolf population is going up, and the beaver population is also going up." One of the students explained that maybe the wolf was killing off a predator of the beaver, and that this could explain the rise in their number of colonies. The teacher replied, "I like your cause-and-effect reasoning here, but you are making some assumptions here without basic facts about these organisms, like their habitats, prey, reproductive cycles. Maybe we can make headway by getting out the information cards about each species—we used these last week, remember?" After the teacher left the group, students pulled out their cards and did some quick reading; they called the teacher back after a few minutes to report that beavers benefited from the increasing willow and cottonwood stands. These plants were beginning a comeback after elk became reluctant to graze on them. Why? Because the wolf packs began patrolling the lowland and riverside habitat where the willow and cottonwood grew.

Most questions that catalyze students' thinking are really just prompts for focusing on one part of a complex story first, so they have an anchor for constructing the rest of the explanation (as with Yellowstone student group 1), or alerting students to relevant information they already have in hand but are not using (Yellowstone student group 2). Every visit you make to students during an activity is unique. You have to think on your feet, even with well-crafted BPQs. But this effort to support sense making is worth it. All it takes is one comment, tailored to students' thinking, to stimulate insights or even breakthroughs in explanations.

We have worked together with hundreds of teachers to identify what makes the "moving among the tables" practice most effective for young learners. The following list reflects our state of the art right now, meaning that this general pattern seems to work better than others to ensure both rigor and equity during the interactions:

1. *Listen first.* Move to the group, and listen to get some sense of the ideas they are using.
2. *Press and point.* Ask questions that either probe students' thinking, or redirect them to some part of the activity or representation that is important to help them reason further (ask about specific parts of their representation or what they are wrestling with, rather than generically asking, "What are you thinking?"). Physically pointing to some part of their representation or a tool they are using is important.
3. *Follow up.* Ask, "What do you mean?" or "Why do you think that?"
4. *Include everyone.* Ask follow-ups to other group members, such as "Do you agree?" or "Want to add on?" Make eye contact with all members of the group. This is a subtle but powerful way to invite students to participate.
5. *Prepare for later share-out (priming).* Determine if the group has a unique idea that should be opened up to the whole class, make a note, and ask if they'd be willing to share. Prep your students about what and how to share in front of the class if they are nervous. Write the group's idea at the bottom or on the back of the BPQ card, so you can remember who agreed to share.
6. *Pose the leaving question.* Encourage students to keep the conversation going after you leave by posing a final question to them.

These rounds are productive for students because the questions you ask focus on their own puzzles, push their initial ideas further, or compel them to voice *how* they know what they think they know. But the rounds also set the stage for a later public conversation about what was learned and how new ideas apply to the anchoring event. The next chapter takes us to that whole-class discussion.

One additional note: if you have not yet referred to the two maps we've created in appendix B, it might be a good time to check them out now. They show where this practice and the other core practices are used during the course of a unit of instruction. We lay out all the lessons that make up two units; one is the fifth-grade unit on sound energy, and the other is the tenth-grade unit on ecosystems we discussed in this chapter. Appendix B provides the bigger picture of how the lessons fit together and when you would use the different core teaching practices.

HOW TO GET STARTED

The most challenging part of this practice is moving among the tables, because you have to be responsive to students' ideas and to who is *not* talking. We advise, for your first try at this, that you select an activity and a written guide for students that clearly have the potential to engage them in reasoning, rather than have them fill out a work sheet. Without this condition in place, the discourse moves you make will fall flat. Write up a set of BPQs, and try them out with just two or three tables. We suggest just two or three because you will find yourself in the midst of sense-making conversations with students and these can take time (10 minutes or more).

During your visits, see if you can get everyone in the group to contribute something to the conversation, even if it is just agreement about what a peer has said. Don't worry so much about getting to all the groups in your classroom during your initial attempts. Focus on the numbered points listed earlier. Try to enact them at the right places and times in the conversation. One issue our teachers encounter in their first attempts is that they tend to press their students with questions, but not use follow-ups. This becomes a steady stream of "how and why" questions that sounds like an interrogation rather than a conversation. We urge you to use the follow-ups to develop students' ideas and soften the interactions.

As you work on this practice, you'll start to internalize the moves; they'll come to you more spontaneously and you'll worry less about mentally "checking off" steps. You'll feel yourself improvising more as you take command of this strategy, recognizing when you are being responsive to students' ideas and getting them to unpack what they say, to compare and contrast claims, and to question each other. As you gain experience, try to engage more small groups in each class. See if you can target these mini-conversations to last about four minutes so that every group will get precious time with you to develop their ideas.

TEN

CORE PRACTICE SET #3

Supporting Ongoing Changes in Thinking: Collective Thinking

WE KNOW FROM the previous chapter that serious sense making can happen during small-group activity. There is, however, another level of talk that you have to facilitate at the whole-class level, in which students take up the next challenge—comparing and contrasting the ways different groups have understood and used science ideas. These public conversations are crucial for learning because they allow students to hear their peers reason in ways that they may never have considered. This is important in a classroom where the goal is to construct better and more defensible explanations over time, rather than focusing exclusively on right answers to work sheet questions. In the whole-group conversation, everyone gets reacquainted with the norms and skills of commenting on the ideas of others, and they experience how uncertainty about science claims is handled when a community builds knowledge together.

The talk, as you might imagine, can get unwieldy. Ideas may erupt from all corners of the room but fail to build upon one another; or, your precocious student in the front row might blurt out a Nobel Prize–winning explanation before anyone else has processed the question you led with. On the other end of the spectrum (or at the same time), entire groups of students may feel they can't or won't participate. And even when students willingly contribute in an organized and equitable way, our own teaching minds are racing through questions like: *Where are we going with this and what's my next move?* For all these reasons, creating predictable routines helps students understand how they can contribute and helps you manage the talk as well as the inevitable dead air.

STRUCTURING WHOLE-CLASS CONVERSATIONS

We recommend that you segment the whole-class discussion into three manageable parts, or mini-conversations about:

- Patterns or trends—what happened in the activity?
- What do we think caused these patterns or observations?
- How does this help us think about our essential question or puzzling phenomenon?

You'd occasionally modify these based on the type of activity students just completed.

Each of these segments has its own goal, which helps students understand the kinds of talk they should offer. In the first conversation, students describe observations or outcomes they recorded during the activity. This is an easy entry into making thinking public because it doesn't require students to explain anything at length or risk making claims. The second conversation builds upon the first—asking students to infer what happened to cause those observations or patterns. The third conversation answers the "so what?" question, getting students to apply what they learned to situations that go beyond the activity itself.

Breaking down the discussion like this keeps you and your students focused on doable talk goals. This talk, however, can disappear into thin air unless it gets documented in an organized way. For this reason, we want students to keep records of the main agreements or disagreements, so they can participate in the moment but also recall days later what was said. For this task, we use a tool called a *Summary Table*. This table can be drawn out on a flip chart or projected as a document from your computer. It typically has a row for each major activity done during a unit (no more than four or five), and three column headings that apply to each activity. These headings reflect the three segments of the conversation we mentioned previously. As you will see in the figures included in this chapter, there have been many variations of this tool created by our teachers, with special features that suit their particular classroom needs.

Table 10.1 shows a completed Summary Table from the tenth-grade unit on wolves in the Yellowstone ecosystem. The teacher used a simple template,

TABLE 10.1 **Summary Table from a tenth-grade biology class studying the reintroduction of wolves to Yellowstone**

ACTIVITY	WHAT WE OBSERVED	WHAT WE LEARNED	HOW IT HELPS US UNDERSTAND THE ECOSYSTEM
Niches–jigsaw reading on coyotes vs. wolves	Some organisms compete for the same space or food. They are adapted to play similar roles in ecosystems.	This is a niche. If the environment they live in changes, it can change which organism is more successful.	The wolf and coyote have overlapping niches. They compete for some of the same kinds of food. When wolves were killed off, coyotes took over. When they came back, the coyotes declined but not all the way because they can eat a wider range of things.
Keystone species and interactions with other populations	Elk decreased in population after wolves introduced. All other species (both animals and plants) increased.	Claim: As wolves increase, it indirectly causes willow to go up. This is because wolves eat elk [and] drive them away from riversides, [and] then willow can grow more. Claim: Bison and elk compete for same grass. When elk get eaten by wolves, bison increase.	Wolves are linked to all other organisms directly or indirectly. Wolves require more energy; they eat mostly elk. Elk is [a] "keystone species" because it is linked to wolves, willows, bison, [and] those crazy beavers, and can also erode river banks.
Conservation of energy and trophic levels	At each trophic level only 10% of the energy got passed to [the] next level.	Energy can get lost as heat and metabolism when you move around a lot. Can't get it back into the ecosystem—its [sic] lost.	There is not room at the top of the pyramid for a lot of predators. So much energy is lost at each level of pyramid that there is room for only a few top predators like wolves or bears.
Carrying capacity	In our "Oh Deer" game we saw that the population of deer decreased when it used up all its recourses, but came back a few years later. Then [it] went down again.	Organisms can only grow in population as long as there are resources for them. They can overshoot their carrying capacity and start to have less [sic] babies because they have less energy to hunt or be healthy. But later the resources come back and so does the organism that depends on them.	All populations in Yellowstone will go up and down in regular cycles, but when [the] ecosystem is disturbed, you can have [the] population crash or skyrocket.
Calories and biomass	Plant-based foods don't create much heat when burned. Fats, proteins burn hotter.	Plants don't contain much stored energy, per kilogram. Fats have molecules that release energy when digested.	Herbivores need to forage all the time to get enough energy to live. Carnivores try to select prey they don't have to chase much to get the energy-rich food.

Good terms or phrases to know: niche, stored energy, conservation of energy, competition, carrying capacity, dynamic equilibrium, claim, evidence and reasoning.

projected on a screen from her laptop. As she managed the whole-class discussions, she had a student volunteer type the final statements into each cell in the row for that day. Notice how the entries in each row require students, as a group, to make deep sense of one or two central ideas. From top to bottom these ideas are: niches; keystone species and their interactions with other populations; trophic levels and conservation of energy; carrying capacity; and calories as storable and transferable energy. Sense making happens as students are deciding what to enter in the "What we learned" column (a variation of "What do we think caused these patterns or observations?"), because they have to use new ideas to interpret the outcomes of the recently completed activity. Sense making also happens as they negotiate the "How it helps us understand the ecosystem" column (in other words, "How does this help us think about our essential question or puzzling phenomenon?"), because they must apply new ideas and relationships to the anchoring event.

Some of our teacher colleagues fill in the lefthand "Activity" column, including the drawings, at the beginning of a unit so that students can see what ideas will be coming up and in what sequence. This encourages some students to read ahead or to predict how the forthcoming science concepts play a role in the anchoring event. English language learners can also translate these key terms before the lessons are enacted, so they are better prepared to participate in conversations.

In broad strokes, here is how a row gets built—more detailed descriptions will follow. When a major activity is completed (investigation, lab exercise, simulation, etc.), you create a new line on the table. Together with your students, you use the column headings as prompts to talk about what they learned and why it matters. You can allow students to take charge of different parts of the conversation or more actively manage it yourself; either way, you'll want to hear from as many students in the room as possible. By "hear from students," we don't mean them just chipping in ideas, but actively commenting on the observations, theories, and puzzlements of others. Gradually, you fill in the row on the table with "bumper sticker" summaries that the whole class feels okay with. There are frequently unexpected insights, contradictions, or confusion about what different groups of students experienced during the activity or what they think. These make your negotiations about what gets written more challenging, but also more valuable for the class to participate in. It is fine to put two conflicting ideas into a cell or put "???" with the intention of filling it in later, when students have more information.

It's important to note that the object of the Summary Table work is *not* to fill in the row as quickly and decisively as possible. The purpose is to have a *sustained conversation* in which students compare and contrast ideas, learn how to build on or critique their peers' reasoning, and apply new science ideas to events that go beyond the activity itself.

Because models and explanations are supposed to evolve over time, and in response to new evidence or arguments, students need to draw upon records of what they've done over the past few days, and to reflect upon what was learned. For these reasons, the Summary Table is one of the most indispensable tools for supporting explanation and modeling. Without some durable representation of what they have done or read, students would have to depend on memory, and each student's memory will have gaps. So, just as scientists do, students keep track of activities and ideas. As the unit progresses, more rows get filled in and, ideally, students start to piece together more coherent and complete explanations by *looking across* the records from different activities in the Summary Table.

Using Summary Tables to Manage Different Parts of These Conversations

Before the whole-class conversation begins, create a new row on the Summary Table by giving the activity a short name (later in the unit you'll be referring to this row by the name you pick, so make it descriptive rather than cute). Have an artistic student do a simple sketch that represents the activity the class just completed. When your students are ready, you can direct them to get out their notebooks and create their own row (not the whole table) with the appropriate headings. When we first started using Summary Tables we found that *students would ask us if it was okay to make their own copy*, to which our teachers enthusiastically said, "Yes!" In most classrooms, we could see that students wanted to add more information to their notebook versions of the table than what we had on the whole-class display. We could see they were making a genuine effort to put into their own words what they saw in the activity, what they felt they learned, and why it mattered to their explanations. This was also a way of keeping more students engaged during the discussion. For these reasons, the notebook entry is now standard procedure in many classrooms.

The first talk prompt you give students is: "So what kinds of patterns or observations did we see?" This question may be different depending upon the type of activity you did (a version of this can also be done with assigned readings). Many of our teachers ask students to come up to the front of the room and share artifacts they produced as part of the lab work (mini-models, flow charts, annotated maps, etc.). Students can even answer questions from their peers about their creations and jump-start peer-to-peer talk that way. Just be aware that this kind of sharing can take a while, so don't spend too much of your "time budget" on this first of three conversations.

We've learned *not* to write the first thing that a student says into the Summary Table row. Rather, you should allow at least two or three contributions, then let students compare and contrast them. Remember to use your discourse moves here. For example, this is a good occasion to press ("What do you mean?" or "What is your evidence?"), encourage peer-to-peer talk ("How is your idea different from what she has just said? And talk to her rather than to me, please"), or revoice ("So what I hear you saying is . . ."). Monitor your own air time. If you do all the talking, then you are the one doing the sense making rather than your students. And stay away from funneling—I know that we sound like a broken record here, but in too many "discussions" we still see teachers trying to get their students to say a specific word or phrase.

Be aware that some students, when asked to talk about *what* the trends and observations were (the first of the three mini-conversations), will instead eagerly begin explaining *why* these patterns emerged. They are prematurely addressing the next column's prompt. Gently ask them to hold off on their interesting insights until the observations and trends part of the conversation is aired out and recorded.

After a few minutes with the trends and observations column, you will want to move on, asking the class: "Based on what we've heard, what do we think is causing these patterns? What can we agree on?" Here, too, you'll have to use your judgment about how to proceed. Often, your students will have some contribution that you feel is too simplistic, or you'll have to write more than one idea in the cell, or you'll have to write a partial idea with a question mark behind it. All of these are fine. Whatever ends up being written in each cell should be in students' language, not in polished textbook prose that you feel is the proper answer (see figure 10.1's entries about the "Human Voices and Vibrations" activity and the "Decibels at a Distance" activity). We give the same

FIGURE 10.1 **Summary Table entries for the "Human Voices and Vibrations" and the "Decibels at a Distance" activities**

Activity	Observations & Patterns	What did we learn?	Connection to the singer?
Human Voices and Vibrations	Whisper • I felt teenie tiny movement in vocal cord • I felt nothing. • No vibration Hum • I felt vibrations • I felt shaking • little vibration Talk • I felt vibrations • The vibration increased Yell • I observed a BIG vibration in my vocal cord. • The vibration got bigger. "Yelling, I felt the most vibration"	• The diaphragm pushes air in and out of the lungs. The diaphragm gets bigger and lungs expands. • air travels from our lungs through our wind pipes into our vocal cords. • There is a vibration in your cords which makes the sound.	The singer used his body to make noise that broke the glass. This is like how we used our diaphragm and vocal cords to hum, talk, and yell.
Decibels at a Distance	purple line: 121 dB @ 2m 99 dB @ 32m green line: 116 dB @ 2m 93 dB @ 32m blue line: 109 dB @ 2m 89 dB @ 32m	- Sound goes in all directions but might not be the same volume - Volume decreases ↓ as distance increases ↑ - air molecules bumping when there is sound	- has to be close so the sound has enough pressure to break the glass - the singer sings to start a chain reaction with air molecules bumping → air molecules closer get bumped harder + ones far away are not moving as hard/strong

advice for the final column about "How does what we learned today help us understand the anchoring event or essential question?"

Let's go back to our time budget. Typically, students can stay focused for fifteen to twenty minutes, total. The final two mini-conversations take the longest to negotiate with students. Thus, it is smart to leave extra minutes for filling in "What is causing these observations?" and "How does this apply to our anchoring event?" In some classrooms teachers have shortened the first mini-conversation (about observations), but in a productive way. Rather than waiting for hesitant volunteers to speak up, they've asked groups of students to write on a sticky note what they observed, then come to the front of the room and place the notes, one on top of the other, on the Summary Table cell. The teacher takes a quick look at the different statements, asking some groups to say more about what they've written. This jump-starts the conversation, and in large classes, it allows every group to have their ideas become part of the Summary Table record.

Another way we've focused the last two conversations is to present opposing opinions from different fictional students. Student A has one idea to contribute and Student B has a different idea (about what caused the patterns/trends, or what the activity has to do with the anchoring event). Everyone in class is likely to feel like they can comment on one or both of the contrasting ideas, and this keeps the conversation from going off in a thousand directions. Of course, we would not want to push this routine too far by constraining novel and generative explanations by students; that would be a big loss for everyone's learning. On the other hand, you will occasionally have to hold students accountable for their statements to be consistent with known science. It's prudent to challenge some conclusions by directing their attention to earlier discussions or readings they've done ("How does your claim fit with the jigsaw articles we read on sound energy?").

Using Summary Charts Instead of Summary Tables for Elementary Students

The Summary Table accumulates a lot of text by the end of a unit, and all these words can overwhelm elementary-aged students. To remedy this, our K–6 teachers often substitute Summary *Charts*, which document students' ideas from only one activity at a time. Each Summary Chart uses the same basic prompts (column headings) as the Summary Table, but students' sense making for a single activity fills the available space.

In a fifth-grade classroom unit on micro-worlds, the anchoring event was a case of food poisoning by salmonella bacteria. Students explored how the microbes interacted with the person's body as they multiplied over time, then declined. Figure 10.2 shows a Summary Chart created by students for an activity focused on bacterial growth. In each of three sections (labeled Observation, Learning, and So What?), students have recorded responses. Clearly there is much more real estate to write and draw on if you dedicate one poster for each activity. Another benefit is that the drawings that represent the activities can, in many cases, be treated like scientific mini-models that stimulate conversations and help students make sense of the phenomenon.

These charts were created by groups of students. The teacher then constructed with students a "consensus version" that was placed alongside individual groups' charts on the classroom walls. In conversations later in this unit,

FIGURE 10.2 **Variation on a Summary Chart for fifth-grade activity on bacterial growth**

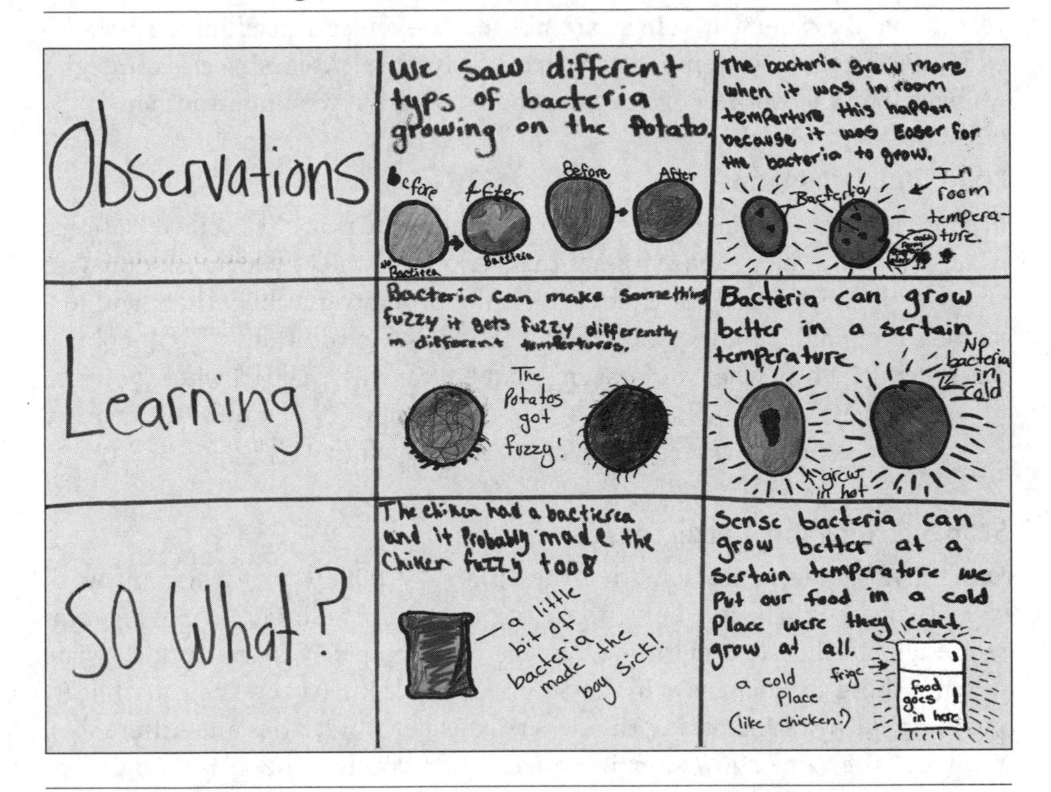

students referred to their own charts more often than the consensus version, suggesting that they felt they had created a useful resource for themselves.

STRATEGIES FOR EQUITY AND ENCOURAGING FULL PARTICIPATION

For all these conversations, you'll want to pay attention to who may *not* be participating and what can be done about it. Students in many classrooms don't feel like they know the rules of joining in when it comes to whole-class discussion. Using the Summary Table provides structure and a talk routine that students can recognize. As you use this tool on a regular basis, more of your students, including English language learners, will recognize how to engage in the conversations. But Summary Tables are not magic—you still need to be proactive about encouraging widespread participation. Next we discuss several strategies we have found to work well, especially in combination with one another.

Point Out Talk Norms

Because students will be referencing each other's ideas in these conversations, ask them to use the sentence frames that are on the wall of your classroom (see chapter 4). They should, for example, use the prompts that help them add to a peer's comment ("Building off ____'s idea, I'd like to add that ____"), politely disagreeing ("I hear what you're saying, but I differ with the part where you said ____, and here's why ____"), asking for clarification ("What did you mean by ____?"), or asking for evidence ("What data tells you that?").

Stop Signaling for "Correctness"

Don't give signals to students that the Summary Table/Chart is a giant work sheet that needs to be filled in with correct answers. These signals can be unintentional, but they do undermine the talk. For example, it is unhelpful to write the first thing a student says in the Summary Table/Chart row instead of hearing multiple views and asking the class to consider similarities and differences. You'll get the same show-stopping effect when you use the terms "right" or "you got it," or even "answer." Once you utter those words, the whole class thinks you heard something you were trying to sniff out and the game is over—no need to add comments to a response the teacher has just labeled as "correct." Frame the activity as a sense-making conversation before it begins.

Prime Groups of Students to Participate

Prime student groups while they are still working on the activity itself, as you move among the tables (see chapter 9). This gives students a chance to debate what they want to say to the whole class later in the Summary Table/Chart conversation, and even rehearse it with your help. Priming helps quiet students contribute and provides opportunities for English language learners to play a role in the presentation of ideas, through talk or drawn representations.

Use "Turn and Talk" Judiciously

Use the simple moves of "turn and talk to your neighbor" when you shift to filling in a new column. This serves a similar function to priming students during small-group work: it gives them a bit more time to compose a response or just to try out an idea with a peer before offering it to the whole class.

These four strategies work together to support wider participation and help students to take risks as they talk to each other about ideas. You might also find it helpful to follow this advice from our teacher colleagues who have successfully used Summary Tables:

- Don't have more than four or five rows (for middle school or high school versions). Choose only the major activities to document on the Summary Table, not every one.
- Students may be able to take charge of negotiating what goes in each column after a reading or activity. At the elementary level, the teacher would take more responsibility for crafting the sentences.
- Don't wait until the end of a unit to fill in the rows for all the activities (we've seen this happen); this is impossible for students. Fill in each row immediately after each activity.
- When students are drawing their final models and writing their explanations, have them use one or two rows on the Summary Table/Chart to identify evidence they are using to support part of that explanation. You don't need students to use the whole Summary Table/Chart and all the evidence expressed within it to support their explanations, but make them accountable for supporting their claims with what they learned through activity and reading.
- Because the completed Summary Table represents activities, new ideas, and the application of these ideas, some teachers create a Summary Table for themselves during the design of a unit. This helps them determine if the combination of activities they are planning includes all the necessary big ideas for students to create final explanations and models. This practice also makes it clearer whether the activities are in a logical order or not.

HOW TO GET STARTED

You can acclimate yourself to whole-class sense-making talk by initially focusing on one science activity and prepping yourself for that conversation. Rehearse in your mind what you'll say during critical moments—for example, when students go silent, when a student wants to spill out the whole explanation using technical vocabulary that others don't have access to, or when two groups of students hold firmly to contradictory interpretations of what they've experienced. Plan a strategy for each of these cases and be ready to use it. Find a private space and say your responses out loud to hear what they sound like.

You'll need to prepare your students, too. On your first attempts, you might script out the words you'll use to frame the conversation to them in order to make everyone as comfortable as possible about contributing. Spend a bit of extra time giving multiple examples of "what counts" as an initial response in each of the columns, and don't forget to model for students the varied ways they can add to, differ with, or challenge a peer's ideas.

Try videotaping your second- or third-period class of the day, after ironing out some of the kinks from period 1 (elementary teachers may not have this option). Alternatively, you can have a colleague observe you trying this out. You can take data on how many students participated in the conversation, or how you responded to students' contributions. It may be informative if you identify how many of the four ways to increase participation (pointing out talk norms, not signaling for correctness, priming students, using turn and talk) you actually used during the class. You and your colleagues can experiment together and then discuss what could be done differently, or identify conditions that seem to help productive talk happen for most students. These "best" conditions may indeed be unique to your classroom.

ELEVEN

SCIENTIFIC ARGUMENT

Making and Justifying Claims in a Science Community

SCIENTISTS ARGUE FOR the same reasons that everyday folks do: to persuade others that something is true or that a particular kind of action should be taken. For a scientist, however, arguments are more structured; they involve making claims, using multiple forms of evidence, and responding to others who critique your ideas. Claims are often about competing explanations for phenomena, but arguments can arise during many other disciplinary activities (see box 11.1 for some sample scenarios). For example, scientists can try to convince others that certain research questions are more fruitful to pursue than others (i.e., they make and defend a claim about this), or that there are gaps in a model and they must be addressed in certain ways. Scientists also argue about methods for collecting data, or how findings from research should be represented back to the science community. Because people can disagree for good reasons about what to think or what action to take, argument, as a disciplinary practice, influences how every part of the knowledge-building enterprise unfolds.

Of all the skills and understandings from the science classroom that students should carry forward into their lives beyond school, none are more important than the abilities to comprehend claims by others, to appreciate the strengths and weaknesses of evidence brought to bear on these, and to deliberate—individually and with others—about what is worth believing or how to take action. These abilities are, in fact, foundational to science literacy. Consider the

BOX 11.1 Situations about which claims can be made

- Why one question is better than another to investigate
- How to design an investigation
- Which tools or strategies work best to collect accurate and relevant data
- How data should be represented for a specific purpose and audience
- Whether a model has a "gap" in its storyline or not
- Whether an inference about what's causing patterns in data is credible or not
- How to revise a hypothesis based on new information

inevitable choices we face, for example, about personal health decisions and the competing claims that surround so many of these. Diet pills promise to subvert our metabolism and help us lose weight effortlessly, but scientific evidence for this is nonexistent and, in many cases, real studies detail the harm that these supplements cause. Similarly, it is rare to hear reasoned arguments in public settings about the alleged links between vaccinations and autism in children, yet some parents forgo immunizations of all types for their children based on little more than rumors or testimonials that cannot be traced back to research of any kind. This puts their families and others in their community at risk of diseases that should not be a threat in modern times. On a global scale, manufactured controversies about the authenticity of climate change are renewed continuously, even by policy makers in our own government.[1] A disturbing proportion of our population (including some science teachers) believe it to be a hoax, despite overwhelming consensus by researchers who have assembled bodies of confirmatory evidence from satellite data of surface temperatures around the earth, historic meteorological trends, studies of shifting ocean chemistries, ice cores from the arctic regions that contain gases trapped from our atmosphere hundreds or thousands of years ago, and the visible impacts on ecosystems worldwide. The diversity and convergence of evidence for human-caused climate change is overwhelming, but somehow our schools, other public institutions, and the media have failed to help large parts of our population conceptualize the mechanisms involved, the magnitude of the problem, and the accelerating pace of change in our biosphere. The point here is that those who want to challenge accepted science should be obligated to understand the nature of evidence they are disputing, how it applies to claims, and what makes assertions from data credible or specious. Children and adults should both learn to play by this clear set of rules.

But where do individuals develop the habits of mind for this? Students may hear arguments of various types at home or among friends, but argument as a science practice is usually explored systematically only in classroom settings. There, students can learn how to frame different kinds of claims, to interpret those of others, to use evidence and logic, to critique and offer feedback, and occasionally to reconcile their claims with competing assertions by their peers—perhaps even agreeing that someone else's ideas are more reasonable than their own. As they attempt to persuade others that a particular claim is credible and listen to the critique of their peers for this position, they come to understand that there is rarely "one right way" of doing things, any more than there are "right answers" to authentic problems in the world.

If you've explored the other chapters in this book, you may wonder: *Aren't different forms of argument already infused throughout Ambitious Science Teaching?* The answer is yes, for the simple reason that so much of students' reasoning is routinely made public, and that they have frequent opportunities to comment on the ideas of others. When we ask students to do things like revise a class list of hypotheses for an anchoring event, each individual's suggested changes can be followed by a teacher prompt: "Why do you think that?" or "What do the rest of you think about the change?" This essentially initiates mini-arguments in which a student has stated, for example, that one hypothesis has more evidence backing it up than others in the list. This is her claim. She is asserting that the list should change and how it should change. The teacher can ask for the student's evidence or for reasoning about how the claim is consistent with known scientific facts. Peers are then invited to offer agreements, critiques, or counterclaims, each of which requires its own justification. These fundamental elements of scientific argument—claim, justification, then critique or counterclaims by others—are embedded in common AST activities such as revising models in the middle of a unit or deliberating about how to design a study that will answer a particular question.

In this chapter we address how to help students use argument to learn. We cannot cover all the varied circumstances in which principled forms of argument may be used, so we will focus on those that begin with claims about explanations (we'll call these *explanatory claims*) because they are so central to AST. We provide examples of these arguments and demonstrate how to scaffold students' attempts at this work.

INTRODUCING STUDENTS TO SCIENTIFIC ARGUMENT

Students learn to argue informally in their everyday lives. They debate with friends about who makes the best pizza in town, or they make a case to their parents that they should have their own phone. But only rarely do students use evidence to back up these claims, and responses by others are often no more than "You are wrong!" (about the pizza) or "That's not a good idea!" (about the phone). So, while students have some familiarity with the idea of argument from outside of school contexts, they can associate it with being right or with emotional confrontations in which someone has to lose.

Because of these preconceived notions, we have to formally reintroduce argument in the classroom as a scientific practice with talk moves like offering claims, evidence, rebuttals, and counterclaims that are used in specialized conversations. But perhaps most importantly, we have to shift students' thinking about the goals of argument from "winning" to clarifying what we (as a group) know, why we believe what we do, and what we should do next.

The basic structure of an invitation to argument is shown in figure 11.1. When we say "invitation," we mean that one participant (or group) uses evidence to support a claim through specific forms of reasoning. Other students can respond to this argument in a variety of ways—for example, by asking for clarifications of the claim or questioning the credibility of the evidence. The original person or group can then respond by rebutting the critiques or simply answering questions posed by peers.

There are far more involved descriptions of what arguments entail, including the use of counterclaims, backing, and other rhetorical maneuvers. However, when too many technical terms are introduced as part of authentic argument, students feel that the talk becomes artificial and they lose the point of the whole activity as well as the focus on science itself. In classrooms, students need only a few moves to engage in vigorous and productive debates.

With these goals in mind, let's look more closely at the key parts of an argument.

FIGURE 11.1 **The initial moves by students that can launch a scientific argument**

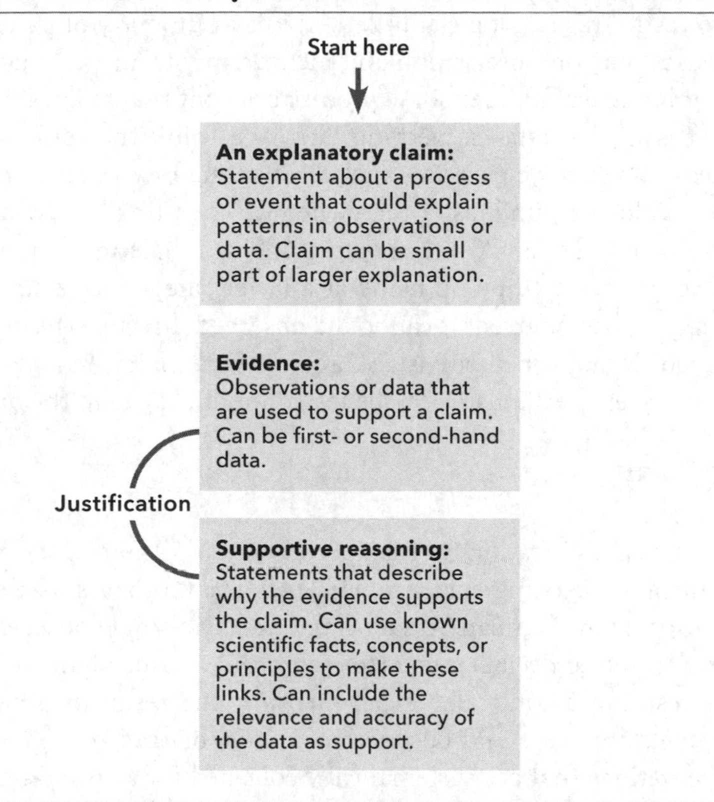

Developing Claims

The first step in argumentation is developing a claim (and remember that we are speaking here only about *explanatory* claims). A claim is usually a statement about some event, process, or relationship in the natural world that the claim's author believes to be true.[2] We can think of a claim as one building block of a larger explanation—but not the whole explanation. A claim is not simply a restatement of trends in data. For example, in a middle school life science activity on respiration, students might record their heart rate and breaths per minute as they move from a resting state to exercise by repeatedly stepping up on a chair and back down. After the data is collected, analyzed, and graphed, the claims we might want them to make are *not* statements like "Our heart rates went up gradually when we started exercising, then went down after we returned to rest." This is not a claim; rather, it is a summary of patterns in the data. It focuses only on observation and measurement. An explanatory claim, on the other hand, would infer some process or event that tells us why we saw those patterns in the data—something like "We think that our muscles are using up oxygen when we exercise, and to get enough we have to inhale more and pump the blood around faster than when we're resting" (there are actually two claims in this statement, one about having to inhale more and the other about the heart having to pump blood at a faster rate). In this statement the "using up oxygen" is inferred; it cannot be observed directly. The requirement to move blood through the body is also a part of the claim. Furthermore, this claim about unseen mechanisms can be tested for "fit" against the data.

Justifying Claims

Citing evidence

A claim requires a justification that can be assessed by others in the class. A key part of justification is citing evidence. Evidence is data that is used to support an explanatory claim. This can be data generated by a single observation, such as when students used decibel meters to record the volume of an air horn. Data can be the result of a controlled experiment, or the result of other kinds of systematic measurements and observations. Students can make observations (data = observations that are systematically collected for a specific purpose) by using one of their senses directly (what they see, hear, feel, smell, or taste) or by using an instrument that extends the senses (a microscope, pH paper to test if something is acidic or basic, a thermometer, satellite imagery, etc.). Remember that *data becomes evidence only when it is used to support a claim.*

Evidence does not have to be firsthand data, meaning data collected by students. It is possible to use secondhand data (i.e., collected by someone else). This was the case with the high school sophomores studying the population trends of organisms in the Yellowstone ecosystem. No student is going to count elk and beaver in our national parks, but they can learn to locate and cite data generated by others to support a claim.

Sometimes when engaging in scientific argument, students have to first discuss whether *the way data was collected* will allow them to see accurate patterns or whether the *right kind of data* was collected in the first place. Perhaps there were not enough observations to draw conclusions, or the data was collected inaccurately. Any of these issues can undermine the credibility of data when students use it as evidence.

Occasionally students can support a claim by using logic and known science ideas *without* invoking new evidence in the form of first- or secondhand data. This will be the case when arguments focus on how to develop investigations or how to display data. Although students can argue about these instances, we want to make sure our students get the most practice stating and reconciling claims about parts of explanations (explanatory claims), using evidence and logic.

Using supportive reasoning

Evidence does not speak for itself; students have to reason publicly with it to make their point. Reasoning is a way of talking about how your evidence supports a claim—it connects the evidence with the claim. This is one of the most challenging conversations you can have in a science classroom. We introduce the basic scaffolding for an invitation to argument in box 11.2, but the examples from classrooms that we describe afterward will make this generic framework easier to understand. Remember that we call this an *invitation* because all the

BOX 11.2 **Basic scaffolding for an invitation to argument**

Our initial claim and justifications

1. Here is our claim [we believe that X is caused by _____ OR we believe that Y has a role in how Z happens].
2. If this claim is true, then when we look at this data we would expect to see [this particular result or this outcome].
3. The reason we'd expect to see this is because [state a brief causal chain of events—this chain has to be consistent with known science ideas/facts].
4. We did see the data pattern we expected. We believe this supports our claim.

Optional additions

5. If our claim were not true, then we'd expect to see [a different set of patterns the data or a particular outcome]. But we didn't see that outcome, so this reasoning also supports our claim.
6. There may be other explanations for the data, such as _____, or _____, but this does not seem likely because _____.

intellectual work described so far is done by a group of students who are deciding how to *begin* a conversation with peers. The other half of scientific argument is the *back-and-forth* between the authors of the claim and their colleagues. Box 11.2 shows a more extended opening than you might expect to hear in the classroom, but we wanted to include a useful structure that students could use to think through their initial reasoning. Within a class setting, students could be challenging each other at any point in these initial statements.

A well-constructed invitation to argument should not have gaps, or places where the audience needs to make big inferences about what happened or why. They shouldn't have to assume too much about events that they don't have direct evidence for.

Looking at a Case of Scientific Argument

Let's explore now an applied example from our fifth-grade unit on the physics of sound. In this scenario, as we've discussed in previous chapters, students were trying to figure out how a singer could break a glass with just the sound from his voice. Early in the unit, some students had hypothesized that sound energy spreads out in all directions and gets weaker the farther it travels from the source. Other students believed that sound travels in waves, but also that it moved in a single direction "where it was pointed" from the source of the sound. The teacher suggested that they go out to the playground with an air horn and decibel meter apps on their phones to see if they could generate evidence for their hypotheses. Students had a hand in designing the investigations and developed a plan for where the air horn should be placed as well as how many phones/decibel meters would be needed and where they should be positioned (these conversations about the design of the investigations could be treated as arguments themselves). Students stood in front of the air horn at 50, 100, and 150 meters. Other students positioned themselves at these same distances to the right, left, and behind the student with the air horn. After they collected data, they created a display to help them reason about their theories (see figure 11.2).

Later, with help from the teacher and after multiple revisions, one group of students used the scaffold in box 11.2 to write and present the following invitation to argument. Their claims and reasoning are not perfect; there are actually two claims being made (about waves moving in all directions *and* weakening over distance), but the argument is credible, including the possible alternative explanation for the data they recorded. This is also just the initial argument by

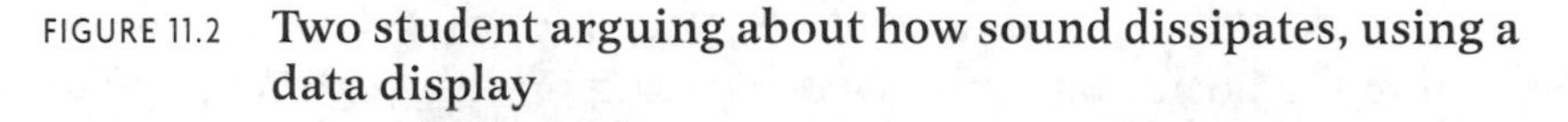

FIGURE 11.2 **Two student arguing about how sound dissipates, using a data display**

one group of students. Other students have to be an audience for these ideas, respond to them, and ideally get into a dialogue with them about the strength of their claims, evidence, and reasoning.

1. *Our claim:* Sound waves go out in all directions from the source at the same speed, and get weaker as they go out.
2. *If this claim is true:* When we did the air horn test, we would expect to see that our decibel meters would show higher readings close to the air horn and lower readings farther away. We would also get sound readings around the whole circle, not just in the direction we were blowing the air horn. We did see this.
3. *Here's the reason we'd expect this:* The air horn is vibrating the air, which makes a wave. Sound doesn't stay in one place. Sound energy has to go out in a circle; that is how sound waves move. Because the same sound energy is spread out over a bigger space, it can't be as powerful in any one spot.

4. *If our claim was not true:* If sound would only go in one direction, then the people behind the horn would not have anything on their decibel meters. But they did record a lot of energy.
5. *An alternative claim:* Some of our friends said that the people behind the horn got their decibel readings because of an echo, but we did not hear any echo and we were out in the open field, so we don't think sound bounced off anything. There were no buildings out there. So, their explanation is not a very good one.

How might peers participate? Figure 11.3 shows a menu of different responses, some of which are questions and some of which are affirmations but with suggestions for peers to reframe claims, cite evidence, or be clearer about the reasoning. Taken together, these responses probe nearly every part of an argument to see if it holds up. We don't recommend asking students to use this entire array—it can be overwhelming. Start with just one or two that are appropriate for your students and expand the choices as the year progresses.

FIGURE 11.3 **Possible responses to argument about explanatory claims**

Is the data open to other interpretations? "When I look at this data I think the trends are different from what you've said . . ."	**Does the claim describe a process or event that may be responsible for trends in the data?** "What do you think causes these trends?"	**Are some observations or data being ignored?** "What are these other observations? Shouldn't these be included when you make your claim?"
Does the claim "fit" with the data? "I agree with your interpretation of the data, but doesn't it lead to a different claim from the one you're making?"	**Does the claim conflict with known science facts?** "Your claim doesn't seem likely because we know from our readings that [state science principle, fact]."	**I agree and . . .** "I agree with what you've said and I think you could also add this evidence to your justification, or you could reason about the claim this way, or you could state your claim this way . . ."
Are there more likely claims than the one you just heard? "I think it is more likely that. . ."	**Does claim only restate the data trends?** "What do you think causes these trends?"	

It's also possible to restructure the conversation so students are comparing two different claims. This sounds more complicated, but it is actually more approachable for novices to argument. All of the basic requirements for the invitation to argument would remain the same, but responses by classmates might now include weighing out the alternative claims and justifications. The bases for comparison would remain the same as for a single claim: the nature of the claim(s), quality and relevance of data, the fit between the claim(s) and known science, and so on. Regardless of whether the class is vetting one claim or comparing two, when these responses are used, the claim's authors should have a chance to think, and perhaps a few minutes to write, before responding. Replies in real time can be intimidating, even for older students.

At the end of a lesson, the teacher needs to ask students to sum up which claims seemed most credible, why, and if anyone had changed their minds in the process (not a forgone conclusion even in the face of robust arguments that contradict one's beliefs).

It is worth noting that we are depicting idealized arguments; they don't really happen in tidy, well-thought-out exchanges. Students will need all possible forms of support for doing this kind of intellectual work. These include the teacher modeling, providing examples and counterexamples of arguments, using discourse guides for talking in these ways, and adhering to norms that maintain civil interactions among students who disagree or have different views about the validity of evidence.

SCAFFOLDING ARGUMENT

Getting students to reason like this is challenging. Table 11.1 organizes for you nine different types of scaffolding that are possible for helping students engage in written and oral argument.

We recommend that you start talking about evidence early in the school year. Take some vignettes from authentic science situations such as the ones we've described in this chapter, where students in these scenarios use evidence to support ideas. Have your class analyze what counted as a claim, what was used as evidence, and how students in the scenario argued with that evidence. You may want to model this kind of talk for students. Be explicit and point out what moves make a good invitation to argument and how to respond. It will take time for students to understand the vocabulary of claims and justification,

TABLE 11.1 **Scaffolds for supporting written and verbal argument**

OFFERING SUPPORTS FOR STUDENTS' WRITING AND TALK	BEING EXPLICIT ABOUT THE LANGUAGE OF ARGUMENT	PROVIDING EXAMPLES
▪ Provide a set of claims about an activity that students have just done, and give them practice at identifying evidence and developing arguments that support or refute one of those claims. ▪ Provide sentence frames to support written and spoken attempts at argument. ▪ Let students create public artifacts or some record of their argumentative reasoning so that other students can learn from them, respond to them, and provide feedback.	▪ Don't assume your students share common understandings of terms like claims, justification, and so on. Be explicit about what they mean and define them. Use these terms in relevant contexts whenever possible and call them out as you use them. ▪ Allow students frequent practice with these ways of using language, perhaps asking, "What claims can be made?" from individual lab activities. ▪ Set up structured times for students to respond to one another's claims in whole-class settings. Provide specific roles for those who are the audience.	▪ Begin the year by using everyday examples of claims and evidence that students can relate and contribute to. ▪ Use authentic but understandable cases of investigations where scientists gathered data to test a claim. Discuss how they gathered relevant data to use as evidence. ▪ Model (think out loud) how you might select a claim to make from an investigative experience, what evidence might be useful, and how to reason about the links.

and to understand the expectations you have for their conversations about explanation and argument.

It is helpful to combine students' experiences in argumentation with their experiences in the design of investigations. For example, you can do a relatively straightforward science demonstration for which there might be different hypotheses generated. You can ask how you might collect data to help test a hypothesis that they favor. After they have collected and analyzed the data, provide sentence frames for students that simply ask them to state a claim based on their investigation, then cite the evidence that they feel supports the claim and why. If it is necessary, you can verbally walk them through your own reasoning about why their claim fits with the evidence, or whether there is a disconnect somewhere in the logic.

We end with an example of an argument scaffold that allows teams of students to give feedback to each other on their claims, use of evidence, and justifications. This can involve everyone in the classroom working on written arguments simultaneously. This example comes from a methods class for preservice science teachers, where we first tested the scaffold. This tool has since been successfully used in many classrooms.

In this course, the "students" are novice educators who participated as learners in the imploding tanker car unit (featuring the gas laws). Near the end of this simulated unit, working in groups of three, they were asked to use the lefthand column of the guide shown in figure 11.4 to draft one claim (just

FIGURE 11.4 **Written guide to support teams, commenting on the arguments of peers**

Goals:

1. To write a scientific claim and support it with relevant forms of evidence.
2. To revise the argument based on feedback from your peers.

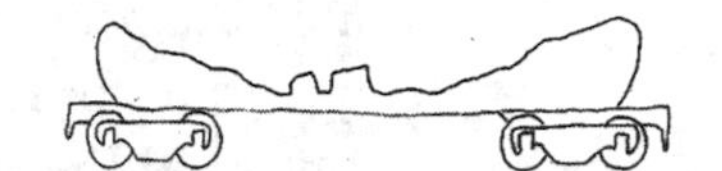

Creating a scientific argument

OUR FIRST DRAFT OF "CLAIM – EVIDENCE – REASONING"	COMMENTS FROM PEERS ON IMPROVING OUR WORK	OUR IMPROVED DRAFT OF "CLAIM – EVIDENCE – REASONING"
CLAIM *Here is our claim* [...we believe that X is caused by...OR we believe that Y has a role in how Z happens...].	Is the claim clear? Is it about what is causing something to happen?	***Revised CLAIM***
EVIDENCE *Our evidence comes from* [name the type of data and the activity it came from]. We saw in the data [name the particular trend or outcome].	Is the data relevant to the claim being made? If two kinds of data or observations are being compared, do they make sense to use together? Is the data credible?	***Revised EVIDENCE***
REASONING *We think this evidence supports our claim because if these trends in data are happening, then it means that* [state a brief causal chain of events–this chain has to be consistent with known science ideas/facts].	Do you need to make big inferences about what happened or why? Are there big gaps in the causal story here? If you saw this kind of data, does it mean that their claim can be the ONLY one that is true? Should they moderate their claim?	***Improved REASONING***

one element in their full explanations), cite relevant evidence for that claim, and provide justification. Sentence frames were provided to help with each construction. Following this, each group passed their guide to another group, whose job it was to understand what the claims and justification were, and provide feedback. These responses are in the middle column. As you can see, there were supports for offering this feedback built into the guide.

About ten minutes into this feedback round, one member of each group visited the peers who were offering them critique. They listened to the group's suggestions for revision and answered questions the group had about the different parts of the argument. When the guide and comments were later returned to the original group, the member who had visited with peers could interpret their suggestions for change. Each group then revised parts of their argument based on feedback from peers. This activity is unusual in that everyone's goal is to *help their peers* develop the most credible arguments possible and to learn to use feedback to improve their own.

HOW TO GET STARTED

The easiest way to introduce students to scientific argument is to have a conversation about everyday situations in which people make claims about food, sports, music, or other topics that young learners are interested in. You can elicit examples of claims that students have heard (or made), then ask what evidence might be appropriate to support these types of claims. You can even get into what makes some kinds of evidence strong (i.e., it's relevant to the claim, based on data, or supports the claim in a clear and unambiguous way) or weak (it's irrelevant to the claim, based only on people's opinions, or supports the claim in an unclear way). After this discussion you could compare and contrast scientific argument with everyday argument. This would be an opportunity to be explicit both about the terminology and how it will be used in class. This is particularly important for your English language learners.

You can then transition from everyday contexts for argument to more academic contexts after students have conducted an activity involving data (first- or secondhand). You can present them with three claims that *you've* constructed, ask groups to select one they feel most comfortable with, and then discuss what evidence from the activity, if any, supports this claim. Selecting from a small menu of claims ensures that students devote their cognitive resources to identifying evidence from the activity that might support the claim and describing how data supports it. The scaffolds we showed earlier could structure their conversations and also allow students to learn from how their peers filled in the sentence frames. You then have an opportunity to share with the class different examples of evidence and justification, perhaps having the authors explain their thinking along the way.

In the past we've used one story that works remarkably well to introduce students to claims and evidence, and in particular the tricky idea of how evidence can be strong or weak, direct or indirect. It's a story about something that happens to your students nearly every day, which captures their interest and makes it powerful for discussion. The story can take different forms, because many of our teachers retell it based on their own experience. We'll share the first part of it, in first person, so you can hear how it sounds:

> I'd like us to have a conversation about evidence today, and what makes different kinds of evidence strong or weak. So here is a story I'd like you to think about. When I was your age, I would come home every day after school, and after a few minutes I could tell if one of my brothers was in the house. Sometimes I would see him and then, of course, I knew he got home before me. But other times, I just had clues. Like I would hear music playing in another room that only he would listen to. That, to me, was evidence that he was in the house. Sometimes I would see his backpack lying on the chair in the kitchen. That was evidence too. I would not see him directly, but knew he was there.

At this point, many students begin connecting to this story, running little thought experiments in their heads. Because they also go home and often get similar clues that a member of their family is already there or not, they realize that they have their own kinds of evidence. We usually ask them to turn to a partner and respond to this prompt: "When you get home from school, who is a person that you might not see, but you can tell that they are there? What's the evidence that they are there?" Our students' responses have been entertaining and varied. They mention things like seeing a sister's coat hanging on a hook behind the front door or noting that their dog had been let out into the back yard, which is something only their father would do. We ask them to use the term *evidence* in these contexts, and students begin to understand what this might mean. The teacher then continues the story, but shifts the conversation to the idea of "strong or weak" evidence, letting students create the criteria for themselves.

With only minor prompting, they reason that, just because one student sees the lights on in her living room, it doesn't mean her mother is home. In other words, there are other people who could be responsible for that phenomenon. Students also agree that some observations can be irrelevant (another good word to understand) to a claim of someone being in the house. For example, the curtains in the living room are left open ("But they are always open!") or the neighbors are parked close to your driveway ("Hey, that's got nothing to do with your family members being home").

Does all this conversation about "who is home" sound hokey? Well, students can roll their eyes when we start in on these kinds of stories, but teachers who have used this example get a lot of responses from young learners who play these thought experiments out, then apply the general principles to the next lab activity. Our colleagues are confident that it's a good way to start a yearlong dialogue about claims, justification, and argument.

TWELVE

CORE PRACTICE SET #4

Drawing Together Evidence-Based Explanations

OVER THE COURSE of a unit, students accumulate many experiences, new language, and new ideas. But stockpiling knowledge is not the goal of Ambitious Science Teaching; instead, we want students to develop more robust understandings by *pulling together* different ideas and bodies of evidence, in order to advance their current explanations and models. These resources include records of their thinking and activity, ideas from their peers, lists of hypotheses, mathematical tools, drawings, metaphors, vocabulary, and, of course, canonical science information. Thus, our title for this core set of practices is a play on words—the "drawing together" refers to students both synthesizing ideas from recent learning experiences, and also inscribing on paper their final models and explanations.

The three practices making up this set, shown in figure 12.1, are as follows:

- co-constructing a "Gotta Have Checklist" with students
- pressing for gapless explanations and models
- assessing for understanding

All three practices are enacted at the end of a unit; however, the practice of pressing for gapless explanations and models is also used once in the middle of a unit. When this practice is used at the midpoint, the goal for students is learning how to *update* hypotheses, models, and explanations based on new ideas.

FIGURE 12.1 **Core practices: Set 4**

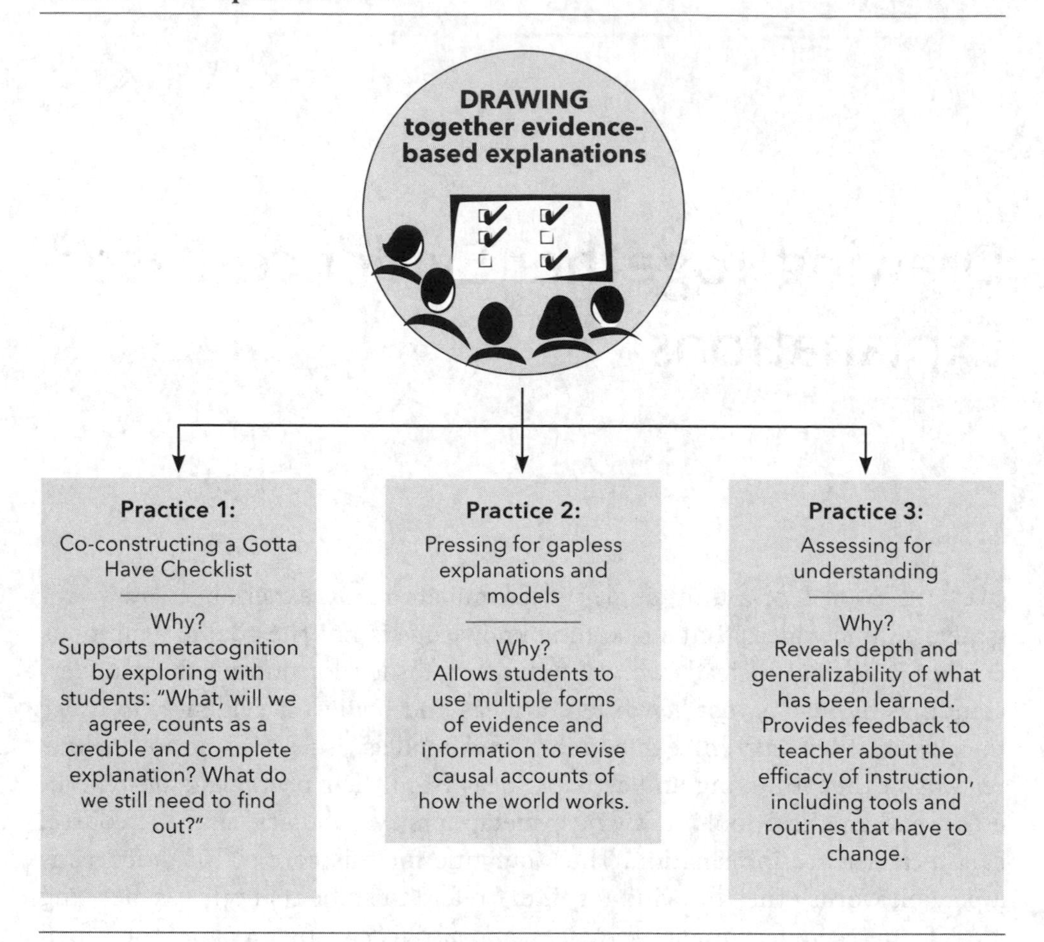

When used near the end of a unit, the practice is aimed at having students *finalize* these same knowledge products. The mid-unit version of these activities acts as a formative assessment of what students currently know, and consequently, what parts of your instruction have been effective or problematic. The end-of-unit version can be part of a summative assessment strategy in which students' products and performances are compared against the target standards.

We'll first describe the version of these practices that happens at the *end* of a unit, then later describe a mid-unit version of the second practice that has fewer "moving parts."

All three practices in this core set share these goals:

- help students learn when and how to use a variety of resources for revising explanations and models
- advance students' abilities to make claims, use different forms of evidence in support of those claims, evaluate claims made by others, and understand how more than one type of claim may be credible in light of available evidence (see also chapter 11)
- support the coherence and clarity of students' explanatory writing (see also appendix G)

PRACTICE 1: CO-CONSTRUCTING A "GOTTA HAVE CHECKLIST"

Over the course of the unit, your students will be exposed to a wide range of science content that could be incorporated into their explanations for the anchoring event. But assembling these ideas into coherent causal narratives is not easy for students; in fact, they are often at a loss for *which* of the many concepts talked about during a unit should be included, and *how* they might fit together. We frequently found that students who had eagerly participated in all the learning activities over the course of a unit struggled to produce a mere two or three sentences, and even those seemed to have random vocabulary terms thrown in—an outcome that didn't reflect these students' state of understanding.[1] To use a medical term, we got "false negatives" about what these students knew. As ambitious educators, we felt we had to own this problem and figure out ways to help our students show more of what they understand.

One solution we discovered, introduced in chapter 7, was for the teacher and students to co-construct a "Gotta Have Checklist"—a set of ideas that had to be included in the final explanations and models (remember, we're describing what this set of practices looks like at the *end* of a unit). Teachers begin, for example, by asking the class, "What are the most important ideas we've explored during this unit? What ideas should really be part of our explanation? We need to make our explanations as complete as possible." With the input of students, you can draft together a set of ideas that most of the class agrees should be part of their explanations or embedded in their models.

You would actively moderate the public construction of the list, using familiar discourse moves (probing, turn-and-talks, follow-up questions, revoicing, wait time). And, because students will often challenge or build on a peer's suggestion, you may want to remind everyone of the norms for commenting on one another's ideas. The Gotta Have Checklist works best when it is *not* a set of vocabulary words; each item on the list should be phrased as an idea. This means that when a student nominates something like "osmosis" to be in the checklist, you should respond, "What *about* osmosis do we have to include in the model or explanations?" In addition, you have to recognize when two ideas on the list are redundant and provide students a chance to combine them or to simply leave one off the list. And finally, if key ideas are not mentioned by students, you must suggest them, or at least prompt students to look back at the Summary Table and ask them, "Should any ideas here be part of our final explanations?"

The final list need not be long at all. Aim for four or five powerful ideas. In our eighth-grade version of the imploding tanker unit (gas laws), we ended up with these four:

Our explanations gotta have:

- ✔ How molecules cause pressure
- ✔ What gases are doing inside and outside the tanker at every phase (before, during, after)
- ✔ How heat energy affects the system
- ✔ How changes in the volume of a container affect pressure

Note that each of these requires students to express a relationship. The first item (how molecules cause pressure) is about unobservable events and links molecular movement with forces on a container wall. Similarly, the fourth item requires students to write about the influence of volume on pressure. We often include the word *how* in our phrasing of the list items because it requires students to write mini-explanations that can then become woven into the larger causal story. Our point here is that the checklist can add rigor to students' explanations and models, and acts as an accountability tool. Without it, your most capable learners may fail to include key ideas, even when they know the science.

The co-creation of a Gotta Have Checklist is a valuable routine, but why would we include it as a *core* practice? Because it invokes generalizable skills that help students reflect on what they know and how they know it. The activity is not just recall for them—it allows them to hear the reasoning of their peers and to interact with each other's ideas.

Our teacher colleagues have developed many productive variations of this practice (remember that all our core teaching practices can be innovated on). Some have incrementally constructed the checklist with students, item by item, over the course of a unit rather than all at once at the end. We have also seen classrooms where students are put in charge of moderating the development of the checklist with their peers. This allows them to take charge of identifying what they are responsible for knowing.

Figure 12.2 is an example of a checklist that a high school biology class produced, moderated by one of the students. The unit topic was how separate species develop from a common ancestor, and the anchoring event was about a line of finches that branched off taxonomically from one another as a result of geographic separation and natural selection processes that were accelerated by a changing environment. The list's items are not stated in their final form (the teacher modified them slightly before students wrote their final explanations), but you can see the class had collectively identified key science ideas

FIGURE 12.2 **Gotta Have Checklist for natural selection**

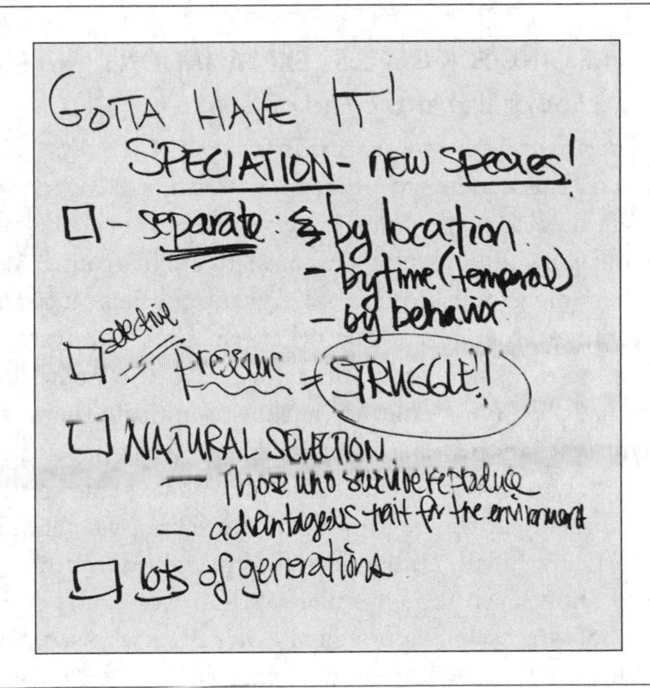

that, together, have the power to explain not only the finches, but many other instances of speciation around the world.

Some teachers have been skeptical about this practice: "Isn't this checklist just a way to give out 'answers'?" Well, none of the items on the lists we've shown here tell students what the ideas mean in the context of the anchoring event, nor how they fit together to form explanations or models. The items simply signal the parts of the explanation or model that have to be elaborated upon by the students. Students are further responsible for stitching all these parts together to create a storyline of the anchoring event over time. We have never seen a checklist enable students who have not made efforts to grasp the science during the unit to suddenly produce coherent causal explanations of complex events. If our critics are still not convinced, we tell them they can do a convenient experiment to test whether the checklist provides an undue advantage. They can provide students with a checklist at the beginning of a unit along with the anchoring event, then see if any of them can draw a coherent model and construct a gapless explanation before instruction begins. No one has taken us up on this challenge yet.

PRACTICE 2: PRESSING FOR GAPLESS EXPLANATIONS AND MODELS

What Practice 2 Looks Like at the End of a Unit

Making critical decisions before students start their work

Now is the time for the students to create final versions of the models and explanations. You will have to make several decisions before this begins. For example, will students work individually? In pairs? In small groups? We tend not to cluster students in big groups (four is a big group) because it becomes too easy for some to dominate the work while others sit back and watch. If you have students in pairs or groups of three, you will want to know who did what, so we have members use a different-colored marker to indicate their respective contributions. If you want individual accountability, you can have students work by themselves. The downside here is that it can intimidate some students to do this kind of task, especially at the beginning of the school year. English language learners are particularly vulnerable here to frustration and not being able to show what they know, even though much of their work can be pictorial representation. Some of our teachers strategically pair ELLs who speak the same first language, and make sure one of them has more advanced English skills.

Another decision you'll have to make involves students' use of evidence to support parts of their explanations and models. Because explanations in AST are so elaborate compared to common forms of school explanations, it's not feasible to ask students to defend every element of them with data and logical argument. We instead ask them to identify one claim within the larger explanation to defend. Sometimes we choose either the before, during, or after part of their models and ask students to use evidence from a class activity to support a relationship they've drawn or written about there.

Yet another decision is about the resources you'll make available to students as they work. We usually allow them to use their notebooks, readings, the Summary Table, and any other public records of the class's thinking (lists of hypotheses, previously drawn models, results of investigations, etc.). In authentic settings for scientists, they always have access to whatever resources are needed. We feel the same about classroom work by students. The catch is that they are not working on simple problems whose answers could be found in a textbook or interpreting the results of a single experiment; instead, students are asked to identify which resources might be relevant, synthesize what these resources provide, and then organize what they know in order to communicate it.

A final planning consideration is about the template that you will provide students. As we wrote about in chapter 7 on modeling, the template for students' *final* versions should include several features that support them in showing what they know. As we describe these, keep in mind the phrase we've used previously: *you won't get what you don't ask for*. This means:

- providing descriptive directions that tell students what ideas you expect to be incorporated (the Gotta Have Checklist may be part of this);
- requesting that they use particular drawing conventions;
- clarifying whether and how they should cite evidence;
- specifying what their models and explanations have to show in each panel (before, during, after); and
- requiring that they represent unobservable events and processes.

Don't forget the obvious: remind students that the model and explanation have to answer the essential question of the unit. We have seen many instances where the essential question was not prominently displayed on the template, and students began drawing and writing about idiosyncratic parts of the phenomenon that failed to represent what they knew.

Finally, the space you provide for written explanations should signal to students how much text is expected. We have found that it helps to have a box—a large box—designated for each part of the explanation (before, during, after). It is even advisable to draw horizontal lines, as on notebook paper, in the box. Believe it or not, this helps some students create more, and more legible, text (see chapter 7).

Moving among the tables for directive critique

If this part of the teaching practice sounds familiar, you are right—it is similar to what you would do during any student activity. After students are given their task you circulate, spending about three minutes or so with each pair or small group, looking at their progress and interacting with them. However, because this is the end of a unit, you offer *more critique* of what students are writing—as they are writing—and are *more directive* in conversations than earlier in the unit. Your discourse moves, for example, should reference the Gotta Have Checklist. After you do a quick analysis of students' in-process writing and drawing, you might select an item from the checklist and say, "Can you show me where on your model you've illustrated this idea?" In our fifth-grade example of the singer breaking the glass with the energy from his voice, one of the items on the checklist was "Show how sound moves through the air from the singer to the glass." At the end of the unit, several pairs of students had drawn waves in their models, like concentric ripples in a pond. The teacher, however, pressed them to show more of what they knew: "I see you have some waves here, but that does not tell me what waves actually are. Take a look at the Summary Charts where we decided that air was made of particles. You need to show what is happening to the air particles—your wave lines just don't tell me enough."

Another common discourse move is focusing students on one part of the model or explanation, pointing to the singer, for example, and saying, "You need to say more about what's happening inside his body before the sound comes out of his mouth." And of course, follow-ups can prompt more targeted writing too: "What makes you say that?" or "Where do you have evidence from our readings or investigations for that statement?" We recommend you bring a pen or pencil in order to point directly to the place on students' drawings or writing where you see gaps—it works better than a finger.

In sum, when you are circulating you are making sure that every student (or pair or small group of students) is showing every relevant thing they know in the form of labeled models and in their writing. If needed, you are prompting them to open their lab notebooks or walk right up to the Summary Table on the wall and have a peer-to-peer discussion about where to find information that is needed for the final product. We cannot stress enough how crucial this version of "moving among the tables" is for student success. It is the difference between coherent models and explanations, and products that are bewilderingly sparse or off-target.

Your critique and prompts are not supposed to ensure that all students' models are carbon copies of one another. In fact, students will choose to emphasize different parts of the phenomenon or make unique parts of the explanation more central than others. This diversity should be encouraged—the only thing that you are reminding students of are the few core ideas that have to be included, regardless of what's highlighted in their models or the angle of the explanations. Students are often proud of their models and explanations. We see fascinating diversity in how they represent their thinking while still including all the important science ideas. We will sometimes do a gallery walk in which all the final examples are put up and class members circulate to visit and learn from each other's work. Occasionally we select a few students to show their constructions under a document camera and narrate their thinking. One of our favorite prompts for students when sharing is "We used to think _____ , but now we think that _____. Here's what changed our thinking _____." Regardless of how you finish your unit, you should take time to celebrate what has been accomplished.

A final note: Writing scientific explanations, especially the gapless explanations typical in AST classrooms, does not come naturally. Some students have skills and tools for doing this demanding work, but nearly everyone needs strategic assistance to get better. This is particularly important for English language learners. For these reasons we've dedicated appendix G to scaffolding explanatory writing.

What Practice 2 Looks Like in the Middle of a Unit

Let's now go backward in time to the middle of the unit. After just a few lessons, students often realize that their initial ideas need revising, or they hear peers advocating in convincing ways for explanations that contradict their own. The middle of a unit, then, is the right time for a special version of the "pressing for gapless explanations and models" practice.

This unfolds in one of two ways. The first option is that students can revise their original models. The second is for the class to collaboratively revise a public list of hypotheses they developed earlier about the anchoring event.

Revising models

In the first revision option, students are given back their initial models and asked to update them. We've learned that students do not like starting their models over from scratch, and only slightly less painful is to draw on top of their initial models or cross out unwanted ideas. Students respond well, however, to putting sticky notes with comments on their initial models.

Figure 12.3 shows the initial model drawn by a fifth-grade student in the sound unit (the singer breaking a glass with his voice), and his sticky-note additions after four days of instruction. His model had been sparsely labeled at the beginning of the unit, including only a few observable features from the video, such as the singer tapping the glass before beginning and the straw moving inside the glass as it started to vibrate. Notably, he also drew musical notes in the second panel.

By the middle of the unit, he now provides evidence that he is changing his thinking about the nature of sound. In the lower left of the model template, he chooses to add an idea; he writes on one sticky note that the "singer's voice create[s] air molicules [sic] around him," and on another note that the vibration caused the straw to move and that the singer's "voice bounced back and forth." In the upper right, he chooses to change one of his original ideas and writes "I would change how I show the singer['s] voice vibrating around the room and glass by changing the repentations [sic] of sound to make it more meaningful because [my initial drawing] just shows music notes that you would see in a music class." He has, in fact, crossed the notes out in the center panel and added vibrational waves.

His revisions tell us two things. One is that his explanation has fundamentally shifted from a simple retelling of what was observable in the video to the idea that vibrations are responsible and that these have some effect on air molecules. He has lingering confusions about people's voices "creating" air molecules themselves (see the sticky note at middle right), but he is aware of his own puzzlement. The other thing that this, and all the other revised models in this class, tells us is which aspects of instruction may have prompted these changes in reasoning, as well as what science ideas the students now need more

FIGURE 12.3 Revised model of singer breaking glass with his voice

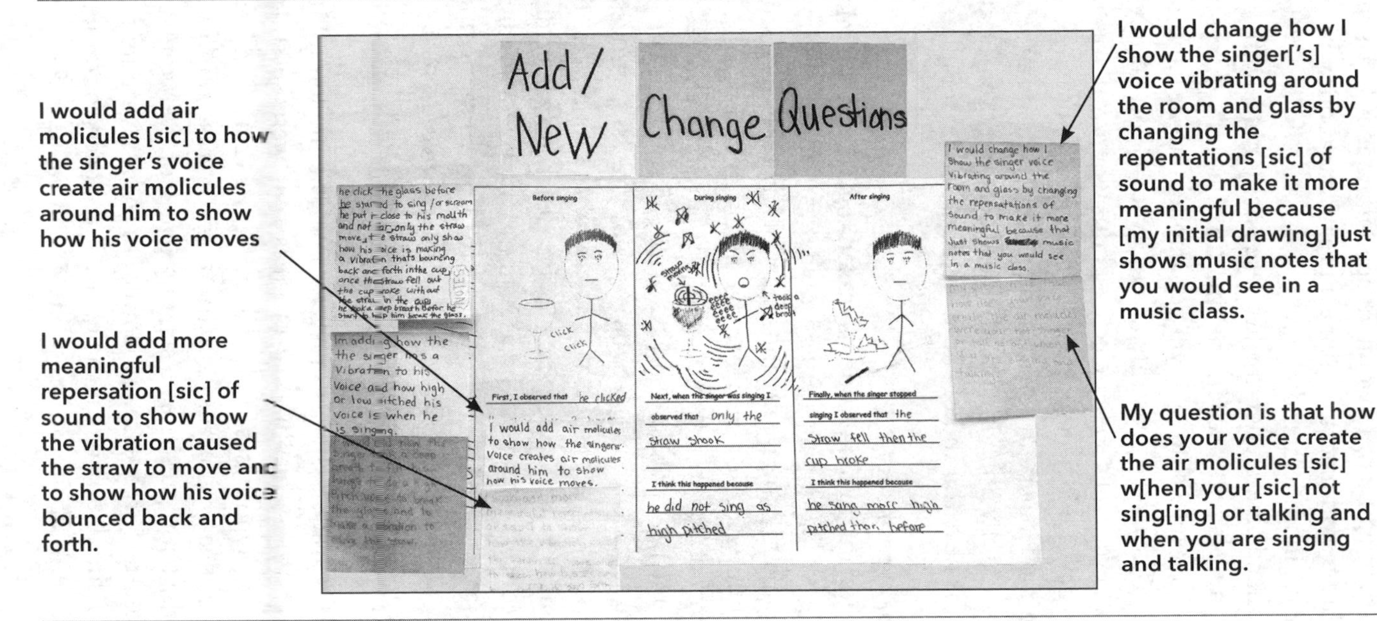

experience with. Imagine what would happen if modeling was *not* a part of this unit. The teacher would have to assume a great deal about what students know or where the gaps in understanding are.

We recommend that students add these sticky notes with commentary to their original models *only once* in the middle of a unit. If you bring out their models every other day, students will get "model fatigue" and be less engaged with the work. They'll be frustrated, too, even if you simply reference the anchoring event every day, so be selective about when you draw their attention back to the big science ideas and puzzles of the unit. Just to be clear: students draw an initial model early in the unit (day 1 or 2), then revise that version in the middle of a unit by applying the sticky notes, and then create final models and explanations near the end of a unit, using a special template you develop.

Revising a list of hypotheses

The second option for helping students reason together in the middle of a unit is to revise the class's list of hypotheses for what caused the anchoring event. We described in chapter 5 how some units start off by recording students' various hypotheses or mini-theories, rather than constructing models.[2] When students co-construct an initial list, it can be populated by partial ideas, simple cause-and-effect statements, or questions. This is a good start, but the list is supposed to change as students learn more. What does this revision process look like? To begin, the list is posted/projected at the front of the room. The teacher presses students to consider how their most recent learning experiences (readings, activities, and conversations) can be used to modify the original ideas. By modify, we mean that: some of the items on the list now seem incomplete and need to be *elaborated upon*; some ideas now seem to be *linked* to other ideas; and still other ideas are *rejected* as implausible or irrelevant, based on science conversations over the past few days.

In our eighth-grade gas laws unit featuring the imploding tanker car, students came up with an original list of ideas that were so simple they could hardly qualify as "hypotheses" (left side of figure 12.4). The items were more like hunches or idea fragments, but the teacher was comfortable recording these because the list represented a starting place for everyone to build from as the unit progressed. Three lessons followed. In the first, students used an interactive computer simulation of gas molecules in a closed container to design and conduct experiments about the questions "What causes pressure?" and "What

FIGURE 12.4 **Initial (left) and revised (right) lists of hypotheses for tanker unit in ninth grade**

Initial list of hypotheses

1. Hatch has to be closed.
2. Tanker can't be filled with anything (We don't know, is it filled with steam?").
3. It has something to do with hot steam.
4. Maybe it has something to do with the day being sunny?
5. Maybe hot air was trying to get out somewhere.

Lesson 1: Computer simulation

Lesson 2: "Stickiness" of water molecules

Lesson 3: Atmospheric pressure

Mid-unit revision of list

1. ~~Hatch has to be closed.~~
2. Tanker can't be filled with anything (We don't know, is it filled with steam?").
3. It has something to do with hot steam.
4. ~~Maybe it has something to do with the day being sunny?~~
5. ~~Maybe hot air was trying to get out somewhere.~~

We want to add these:

- *Open hatch*: If the tanker is open then the air molecules can move in and out of the system, making the pressure inside the tanker equal the pressure outside of the tanker.
- *Energy transfer*: If hatch is closed, lots of the same molecules inside the tanker hit the interior. Some of that energy is transferred to the wall; they bounce back with less energy. If the water molecules lose energy, they are more likely to stick to each other because they are attracted to each other.
- *Slow down and sticky*: When water molecules slow down and stick to one another, it has two effects: one is that they are moving more slowly and hitting the walls with less energy, and [the other is] that liquid is condensing and their droplets are falling to the bottom of the tanker, and no longer hitting the walls.

conditions can increase or decrease gas pressure in a system?" In a follow-up lesson, students discussed a reading that illustrated how all water molecules in the air had positively and negatively charged sides to them, and that because opposites attract, the molecules could stick together (i.e., lead to condensation). Later that lesson, with some teacher focusing, students theorized that this sticking together could happen only if the water molecules slowed down, because normally they move too fast and simply fly past one another. This led to questions about what might cause water molecules to lose speed in a system. The third lesson involved a lab on atmospheric pressure, in which students learned that they were surrounded by an "ocean of air." This air exerted fourteen pounds per square inch of pressure on every surface at sea level. Some students then suggested comparisons to submarines that are built to withstand water pressure from all sides. This, in turn, catalyzed a conversation about whether the tanker was built to withstand pressure from inside or outside, and whether it mattered where the force was coming from.

So, only four days into this unit, students were ready to revise their list of rough hypotheses. The teacher placed them in groups of three and gave them these prompts:

- Select one hypothesis or hunch from our original list.
- Decide whether you should (1) get rid of it, (2) elaborate on the idea to make it more consistent with what we know now, or (3) connect it to another hypothesis or hunch.
- Should we add any new hypotheses?
- Should we put these in order of when we think they happened?

Then, after a reminder to look back at the Summary Table entries they had made for the three previous lessons and five minutes to chat in their small groups, students were ready to propose changes to the list. The teacher asked one member of the class to step up to the computer to type agreed-upon changes into the projected list. From that point, students did most of the talking. As we can see from the right side of figure 12.4, students' ideas were changing—not from wrong to right, but from small grain-sized ideas dealing with observable phenomena (steam, heat, the hatch, a sunny day) to hypotheses about theoretical events and processes (energy transfer, pressure inside and outside, molecules sticking to one another) and statements that combined ideas together. They kept items #2 and #3 from the original list and crossed out the rest.[3] These two survivors now seem simple compared to the three bulleted additions, but the aim here is to improve and expand upon ideas, rather than making the list look neat. The teacher asked students to name each of the additions (for example, "Open hatch") so they could easily refer to it when questioning what it meant or offering friendly amendments to it.

The items on the list now are starting to look like parts of a coherent explanation. At a glance, the teacher could tell where the class was making progress and where challenges still lurked. Some students, however, did not participate in the list revision, so the teacher asked everyone to fill out an exit ticket with two questions: Which ideas that we talked about today were still confusing to you? What information do you still need to help you understand the tanker better? It is easy to get the false impression after an inspiring group conversation that everyone understands what was being talked about.

When you moderate such changes in a list, you can use familiar AST moves to help students participate—for example, by framing the activity as authentic science work, providing suggestions for what one can do with ideas on the list (elaborate on it, link it to another idea, cross it out), and using turn-and-talks. You can use combinations of discourse moves like pressing, follow-ups, and wait time. But most importantly you should be prompting cross-talk among your students. Try to have them interact with their peers, to address each other directly rather than talk through you. Move to a corner of the room if you have to. We've also seen students moderate this discussion, rather than the teacher. With just a little help from you, they can be surprisingly effective at getting their peers to open up. They take great pride in running the show, and these student-led discussions can create a sense of community in which they feel some agency about which ideas get aired and how.

Regardless of who moderates the revision of the list of hypotheses, students have to be held accountable for their suggestions to be consistent with science. Someone has to occasionally ask, "What reading or activity makes you think that?" or "I hear this claim being made; does anyone agree or have a different idea?" This may be a good time to talk about evidence from multiple experiences. If someone insists on a revision that we know is outright contrary with known science, we usually put a question mark in parentheses (?) after it. This isn't to be coy or to signal that it's wrong; it's for the teacher to exercise responsibility in questioning a statement that does not have strong evidence behind it. By the end of the revising session, the list will look more like an outline for a full explanation and less like an eclectic assortment of intuitive ideas.

While revising the list, you might want to add the element of scientific argument by treating the hypotheses like claims that have to be supported. Students can use specific criteria to support or refute particular statements on the list. Where do these criteria come from? In one of our sophomore biology classes, the students and teacher co-created the "How good is your hypothesis?" poster shown in box 12.1. Using a set of guidelines like this makes criteria explicit for authentic disciplinary argument.

Students referred to these four criteria all throughout the unit to get conversations going. By the way, a hypothesis in this case is very much like a claim.

BOX 12.1 Student-created criteria for assessing hypotheses

How good is your hypothesis?

"The likelihood test"

How likely is it that the events in your hypothesis can actually occur? Are there parts of your hypothesis that depend on really unlikely things happening?

"Fits with all data patterns"

Can the hypothesis explain most or all data patterns?

"Based on known science"

Is the hypothesis itself based on well-known science knowledge like facts, laws, and theories? Or is it contrary to basic facts and knowledge?

"Based on reliable data"

Is the hypothesis based on data that really measures what is important? Is the data reliable and credible?

We realize that there are official ways that these terms are used in science, but for learning situations in classrooms the distinctions do not always matter.

PRACTICE 3: ASSESSING FOR UNDERSTANDING

In AST we think of assessment as a system used over the course of a unit to accomplish three goals: improve instruction; provide feedback to students on their current levels of understanding; and make final evaluations of student learning.

Our teachers work daily on the first two of these goals by using a repertoire of formative assessment strategies. These strategies can be so woven into the fabric of AST that they are invisible to educators who are new to this work. For example, eliciting students' ideas at the outset of a unit gives you a picture of what knowledge students are starting with. Later on, all the modeling work, the Summary Table discussions, exit tickets, the small-group conversations as you move among the tables—all of these are assessments that make student thinking visible, and are opportunities to provide feedback to students on their thinking. In addition, they provide data you can use to guide instructional decisions; much of it tells you in clear terms what parts of your instruction were effective and what needs to change or be supplemented. Importantly, these formative assessment strategies also tell you which groups of students are not benefiting from your instruction in the ways you intend.

On the other side of the coin, evaluative (or summative) assessments tell us what students have learned and allow us to provide grades. Performances by students (modeling, explanation, argument, designing or conducting investigations, etc.) can be treated as part of a larger summative assessment strategy, when combined with more traditional tests that have questions and prompts. To guide the design of summative assessments, we use a set of principles rather than teaching practices. These suggestions are not meant to be a comprehensive guide to assessment practices; rather, they can help you think about a repertoire of strategies, relevant to AST units, in which the aim is for you to document the types of intellectual work your students can do.

Principle 1: Assess What Was Taught

This seems like a no-brainer, but students often take quizzes and tests that have been downloaded or recycled from the file cabinet. This makes school seem like a game in which you can be held accountable for things you never studied. Here's the simple rule: all learning activities should be assessed in some way, and conversely students should not be given assessment tasks or items for which there were no corresponding learning experiences.

Principle 2: Use Authentic Assessment Tasks

Authentic assessment tasks simulate intellectual work that is valued in real-world contexts. These require students to use ideas and skills, for example, to engineer solutions for problems, interpret the results of investigations, or communicate with specific audiences about scientific information. Throughout our units we usually provide students with realistic scenarios in which these kinds of goals have to be achieved and there are certain resources to do the work (information, background context, etc.). We combine these tasks with more traditional question formats, like multiple-choice (used sparingly) and short or long open-response items, to get a more complete picture of what students know.

Principle 3: Make Criteria for Success Clear to Students

Any student performances should be guided by explicit criteria used to judge its quality. This is, for example, what the Gotta Have Checklist was designed for. The checklist can become the basis of a grading rubric for students' models and explanations; you just have to unpack for them what characterizes acceptable and exemplary responses for each item. Occasionally you'll want to assess students' abilities to formulate claims and use evidence to support those claims. If, however, you simply direct students to "include evidence" you may just get what you asked for—references to data or observations without any reasoning that links it to their claims. Technically, they've met your criteria. If students don't show their reasoning or they show it in a way that doesn't match your expectations, that could be your problem and not theirs. As your assessment tasks become more authentic and complex, students will need more guidance about what constitutes high-quality work.

Principle 4: Use Combinations of Lower- and Higher-Cognitive-Demand Items

One category of high-cognitive-demand task is to ask students to apply science ideas they've learned to new situations. For example, in the final assessment for the sophomore biology unit on the reintroduction of wolves to Yellowstone, we provided students with climate change data showing that increasing temperatures were allowing certain non-native insect species to move north into the ecosystem (yes, it's really happening). We provided information about the insect species and then asked students to predict population shifts in two other organisms, native to the park, and support these predictions based on what they knew about species interdependence. This is a "what-if" scenario in which one or more elements of a familiar system are changed. Students, hopefully, use what they've learned during the unit to explain what they think will happen and why.

This is also an example of a *near-transfer* task. Students apply knowledge to situations that are very similar to those they've already studied. *Far-transfer* tasks are more challenging because they require students to use knowledge and skills in situations that bear less apparent resemblance to those they have studied. In the fifth-grade unit on the singer breaking the glass with the energy from his voice, students could be asked, for example, to model how a loud car stereo can cause windows on nearby vehicles to vibrate. In these assessment tasks, students must recognize the phenomena (sound energy traveling through space) as similar to what they've already studied, and that the causes (human voice versus stereo speaker) and effects (wine glass shattering versus car windows thumping), though different, are related by the same unseen events and processes (compression of air particles against one another, wave motion through three-dimensional space, resonance).

Because near-transfer and far-transfer tasks are both challenging for students, you wouldn't want them to face several of these on a final assessment. Not only would that be grueling for them, but it also would distort what you aim to measure. It's possible, for example, that some of your students could not complete the more demanding application (transfer) tasks to earn even partial credit. This means that you would have no way of knowing whether they were confused by higher-cognitive-demand requirements of the assessment or whether they did not grasp one particular science idea needed to simply begin the task. If our fifth graders could not begin to model the car stereo phenomenon, they may have failed to understand only the idea of wave frequency being a function of how many times the speaker moved back and forth per second. They may well have understood all the other ideas needed for the task—like how sound energy travels in waves, the nature of compression waves, and the idea of resonance. Not knowing how a car speaker operates would cost them a lot in this case, but looking at their responses, you would never know that this was the only thing holding them back.

There are solutions for this. If we want a more complete picture of what students know, then we must have some basic, lower-cognitive-demand question and tasks on our assessments that separate out selected science ideas. In the high school ecosystem unit, for example, our teachers have included individual items about the ideas of carrying capacity, how energy is passed from producers through different levels of consumers, and how social behaviors of species like wolves and beavers can enhance their chances for survival. These are asked outside the context of the more complex modeling scenarios.

Assessments then, especially summative, should be composed of a mix of items and tasks with differing levels of cognitive demand. The items with low cognitive demand should be able to tell you whether students understand some of the key ideas needed to contribute to success but struggle on the more contextualized and higher-cognitive-demand tasks.

Principle 5: Provide Equitable Opportunities for Students to Show What They Know

Some of your students will need special forms of assistance to complete your assessments. Some will need more time, and others will need directions read to them, questions written in their native language, illustrations that accompany the items, or the option to talk out rather than write responses. Students may need help understanding which activities or readings they have done are relevant to the assessment tasks. Many of these forms of assistance allow you to maintain the full rigor of the assessment questions and tasks—all you have to do is remove barriers that prevent students from showing what they can do. Even if parts of the assessment have to be modified for some students so that questions are made more basic, this still allows them to let you know what they've learned. As an alternative for students who are particularly disadvantaged by conventional assessments or have test anxiety, we have had them produce a portfolio of artifacts (models, changes in hypotheses, accounts of investigations written in lab notebooks, reflections on what they've learned from readings, etc.) in which they accumulate records of work and thinking over the course of the unit.

HOW TO GET STARTED

Try your hand at the Gotta Have Checklist. Generate with students four or five items during a lesson near the end of a unit, but make sure you later massage the wording and add any ideas that seem to be missing. The trick will be constraining the list to fewer than a half-dozen ideas. Any more than that, and you are likely asking students for nonessential details rather than core explanatory concepts.

After students complete their final models and explanations, you can analyze a sample of them to gauge how effective the checklist was in supporting them to show what they know. Did students seem to ignore the items and create explanations that had no clear link to the checklist? If this is the case, you may not have been assertive enough as you circulated among the tables. While they are in the writing/drawing process, you have to ask every student or group of students to show you where at least one item on the list is represented in their work. But let's say you were careful to make these requests and their models and explanations still seem incomplete; perhaps you failed to include a key idea in the checklist. Often our teachers find that they've neglected a core concept, and the oversight becomes apparent when different highly capable students leave the same gaping holes in their explanatory storylines.

Another reason students leave out particular science ideas is that they simply don't understand them well, a problem that can be traced back to their opportunities to learn during the unit. The obvious fix is to give students more sense-making time on those ideas during the unit, but your assessment system should help you determine if you are working on the right instructional issue. This is why we advocate for summative assessments that include questions and tasks that go beyond the modeling and explanation to probe students' understanding of individual concepts. These items help you see whether students' misunderstandings of key ideas are the likely reason they can't complete more complex tasks that require this basic knowledge to build upon. For example, if your seventh graders can't explain the regularity of solar eclipses, it could be because they don't realize that gravity plays a central role in the predictable motions of bodies in our solar system. Without that key idea, their larger explanations can't get off the ground.

Occasionally, you'll have done everything according to plan—designed meaningful learning activities for all the big ideas in the unit, constructed with students a thoughtful but compact Gotta Have Checklist, and circulated while they were working and requested to see how checklist items were reflected in their writing. But somehow your students still struggled to produce a narrative for the anchoring event that is coherent and complete. If this is the case, they may not be challenged by the science itself but by the nature of explanation writing. In that event, they simply need opportunities to rehearse with some scaffolding and get feedback from you on small-scale attempts before being confronted by high-stakes, summative assessment situations.

THIRTEEN

Organizing with Colleagues to Improve Teaching

TEACHERS WHO SUCCESSFULLY change how their students learn science most often share the risks and challenges of innovation with colleagues. In best-case scenarios, diverse teams of educators in a school, or even across schools, come together regularly, over months or years, to work on the systematic improvement of instruction. Principals who take their roles as instructional leaders seriously are becoming increasingly supportive of such efforts, setting aside regular times for faculty to meet and giving them the latitude to set an agenda for experimentation and change in their classrooms. These professional arrangements, however, require an "infrastructure" to be productive. This includes shared goals, protocols that guide how the team makes decisions together, and tools for inquiry into practice. Without this support, it is too easy to repurpose meeting time for common planning, passing along news about students, or examining new curriculum materials. In this chapter, we draw upon lessons learned from partnering with over three hundred K–12 teachers in local professional learning communities (PLCs) to improve their classroom practice and reshape the conditions under which students learn science. All these communities have focused on the same three questions:

- Which practices work best in *our* classrooms?
- Under *which* conditions?
- And for *whom*?

These prompts have helped teachers adapt ambitious practices to their local contexts and probe issues of inequity that had been "flying under the radar" in their schools.

Our vision for improvement work is informed by the Carnegie Foundation for the Advancement of Teaching and its strategies for empowering teachers to drive change for themselves.[1] Central to this work are *Plan-Do-Study-Act* (PDSA) cycles, which involve rapidly studying small-scale tests of instructional change in a couple of classrooms (such as adapting a discussion strategy to better meet the needs of students, or introducing a new tool to support explanatory writing), and, if successful, moving to broader-scale tests of this innovation. "Broader-scale" might mean implementing the practice or tool in all science classrooms of a school, or integrating it as a regularly occurring routine in each classroom. The PDSA model of improvement is illustrated in figure 13.1, but we acknowledge that this represents an "ideal" and cannot fairly depict the challenges and opportunities PLCs encounter as teachers work together on improvement.

FIGURE 13.1 **Cumulative learning through Plan-Do-Study-Act cycles**

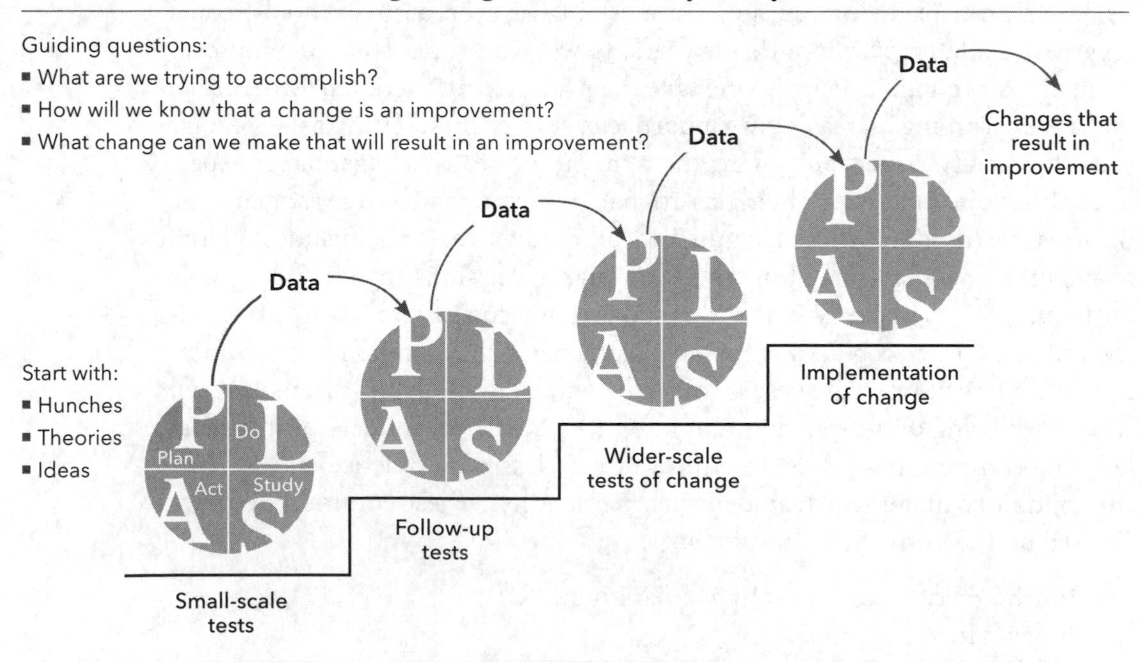

Progress is not always a steady stream of successes, but even failed attempts at change are a part of learning, especially when teams keep focused on their larger goals. And the risks we mentioned earlier? Most teachers would agree that deprivatizing their practice makes them the most apprehensive. After all, one's professional identity is on the line when peers file into your classroom, notepads in hand, to watch how you work with students.

Teachers can form PLCs within or across grade levels; there are benefits to each option. PLCs with teachers from different grade levels can work to coordinate students' learning experiences across their school science careers. For example, team members could be purposeful about helping students engage with increasingly sophisticated versions of scientific modeling, explanation, or argument over successive school years. On the other hand, teams of same-grade teachers could experiment with instruction about particularly challenging science ideas or disciplinary activities that are common to their curricula, and have the option of testing changes to practices or tools in multiple classrooms at once.

Regardless of how your team is formed, it is wise to invite instructional coaches—perhaps ELL or science specialists from the school or district—and assistant principals or principals to participate in the PLC meetings. These individuals can offer unique perspectives on what's happening in classrooms, identify which students are not being served well, and suggest approaches that might be worth trying. Based on our research, we suggest that teams meet at least once a month to make progress on PDSA cycles. We organize the remainder of this chapter by the different phases of the Plan-Do-Study-Act cycle, and finish with some tips on getting buy-in from administrators for this work. First, however, we describe crucial foundations for the team's efforts.

LAYING THE GROUNDWORK FOR PDSA CYCLES

Choosing a Facilitator

Any group of professionals seeking to do improvement work needs a big-picture person who empowers individual team members to contribute and makes sure these individuals function well together over the long term. Good facilitators set the stage for rich conversations by managing time, delegating responsibility when necessary, keeping the group focused on important goals, and fostering a sense of both collegiality and accountability. Facilitators can be senior members of a science department, a district coach, or any science teacher who has the motivation and skills to listen actively and to press for change. They set up meetings and communicate with teachers and administrators about the PDSA cycles. We've found it helpful for the facilitator to work with school leadership to secure times for teams to learn together on a monthly basis, and to work with the team to decide how AST practices support schoolwide goals.

This coordination of goals is important. Principals often have their own broad instructional initiatives that they want to see applied across subject matter areas, such as emphases on reading and writing, formative assessment, or standards-based teaching.[2] We have seen cases where a team's improvement goals have been sidetracked by these general aims, because science teachers could not articulate the value to administrators of the discipline-specific work they wanted students to do. In other cases, team members could not envision how these broader initiatives could be taken up as part of AST practices. We suggest that coaches or teacher leaders take instructional walks with principals to help the leadership understand AST practices and what student learning looks like at the beginning, middle, and end of a unit of instruction.

While the team is working on improving instruction, facilitators can work on improving their *own* skills by monitoring and asking for feedback on the following three clusters of facilitation practices:

- *Promoting full participation.* Encouraging participation and engaging the group in monitoring all members' participation, inviting others to share their expertise and take on roles within the group, validating others' ideas, and standing back when necessary to let others participate.
- *Staying close to data.* Maintaining a focus during meetings on data and observations by asking others to talk about evidence that indicates problems or breakthroughs in teaching, and to provide samples of student work to assess thinking or growth.
- *Sustaining inquiry into learning.* Sustaining the team's inquiry stance about the relationship between teaching routines and student learning, offering conjectures about student learning for the team to discuss, pressing team members to articulate connections between their classroom activity and student learning, and countering others' explanations, when necessary, by offering alternative cause-and-effect stories about student learning.

These responsibilities may seem daunting at first, but everyone on the team should understand and work on these together. It's not just up to the facilitator.

From a broader perspective, PLC work should be synchronized with other kinds of development efforts for the science team. For example, teams make better progress when various types of professional development opportunities do not pull them in multiple directions. In one school district we've partnered with, all secondary schools in the district coordinate one-on-one coaching, monthly PLC time, summer workshops, instructional walks with principals, and Studio Days (explained later in this chapter) so that they all "push in the same direction."

Setting Professional Learning Community Norms

Co-constructing team norms is an important first step. All participants should weigh in on at least six dimensions for productive joint work. We've phrased these as questions you might pose in your first meeting:

- How much time should we devote to PLC meetings and how do we start and end on time?
- What are the expectations for individual responsibility to the group?
- How can we listen to one another and encourage everyone's participation?
- How will we honor risk taking and confidentiality?
- How should we make decisions, work toward agreements, and deal with tensions?
- How will we stay focused on students' ideas and experiences as the object of common work and inquiry?

You shouldn't expect consensus on all these questions, but highlighting such issues up front makes the group better prepared to handle tensions that arise as they do this work over time. Teams can use protocols to support the writing, enforcing, and routine revisiting of group norms.[3]

Team members should be particularly vigilant about norms related to the topics of conversation at meetings. For example, participants will want to adopt an inquiry stance toward students' learning (e.g., how and why students respond to instructional situations, special supports, or their peers) rather than engage in "repair talk." Repair talk can crop up frequently; it sounds like recommendations for what should be done, or what will "fix" a problem. These comments originate from a generous place and often draw upon one's own personal history of experiences, but the instinct to help can foreclose on everyone's efforts to understand what is happening in classrooms.

Selecting an Improvement Focus

Prior to the school year, teams should survey their commonly shared practices and tools, considering which of these might be improvable and thus serve as a focal point for instructional experimentation. Defining a practice to work on can be challenging since teachers typically do not publicly unpack how they teach or even think of the complex work they do every day as a set of discrete practices.

Teams can select a focus such as pressing students for evidence-based explanations, but they should also select a specific aspect of that practice (a subpractice) that can be worked on more directly. The teams we've partnered with have worked on five subpractices, each of which is a smaller grain-sized activity, representing a specific opportunity to learn for students:

- Using structured partner talk for "how" and "why" reasoning
- Sequencing share-outs of students' initial models
- Using peer feedback to deepen written explanations
- Using language functions as a lens for reading, writing, and modeling
- Revising lists of student-generated hypotheses with evidence

Most teams that are new to improvement work select structured talk for "how" and "why" reasoning (see appendix D) or sequencing share-outs of models (chapter 7). Whatever subpractices you choose, they should play some central role in facilitating students' reasoning, writing, or participation in a valued activity.

The relationships among the subpractices and the overall team goal are shown in figure 13.2. This is called a *driver diagram*; it illustrates the relations between team goals and teaching practices—in three layers. First, it specifies one high-value goal for the PLC work. Second, it represents practices that impact the achievement of that overall goal (these are called *primary drivers*). Third, it shows the subpractices (*secondary drivers*) that the team believes most directly influence students' opportunities to learn and ultimately help attain the goals at the top of the diagram. We refer to these secondary drivers, or subpractices, as "actionable" because they are what the team can improve most directly. A driver diagram is a team product—it requires multiple meetings and

FIGURE 13.2 **Driver diagram specifying links among goal, primary drivers, and secondary drivers**

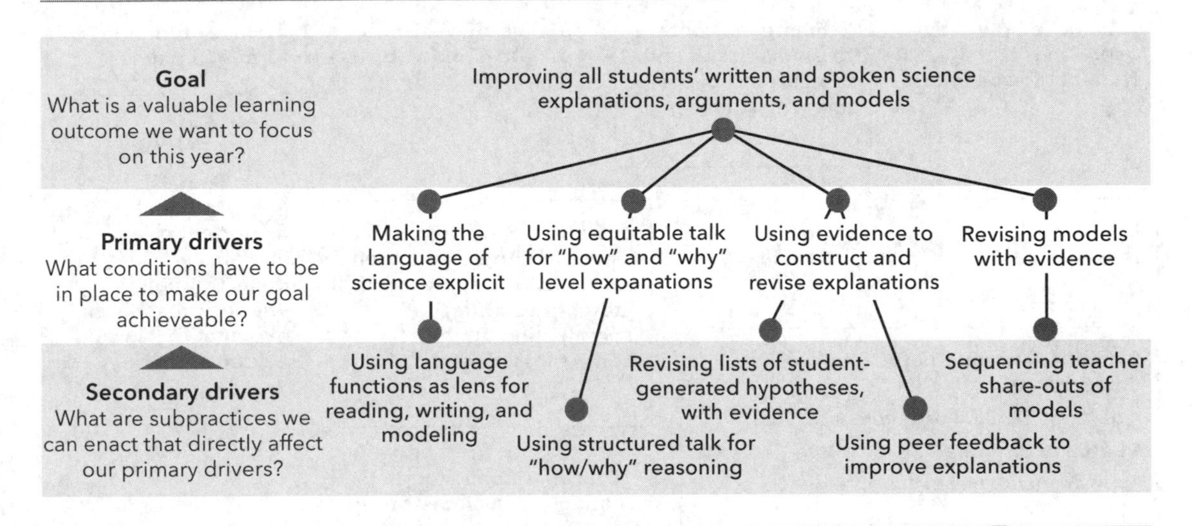

negotiation with colleagues about what is most important to improve in the classroom. Importantly, the lower parts of the diagram are largely about things you'd see and hear teachers doing with students. The subpractices are observable, so you can specify them as a team and collect data about how they unfold during instruction.

Getting Clear About the Focus of Change

After your team has created the driver diagram, the next step is selecting a particular practice, or more specifically, a subpractice you want to improve: What is it? What purpose does it serve? How does it work? One way to start answering these questions is for the team to define a *typical sequence* for the subpractice, a list of non-negotiables for enacting it, and a tool that students can use during class to support their learning. Figure 13.3 shows a possible teaching sequence for structured talk for "how" and "why" reasoning—a subpractice that is intended to support ELLs in developing fluency with both English and the academic language of science, but supports *all* students in deepening their reasoning about a phenomenon (see chapter 4 for a refresher on structured talk).

FIGURE 13.3 Typical teaching sequence for structured talk for "how" and "why" reasoning

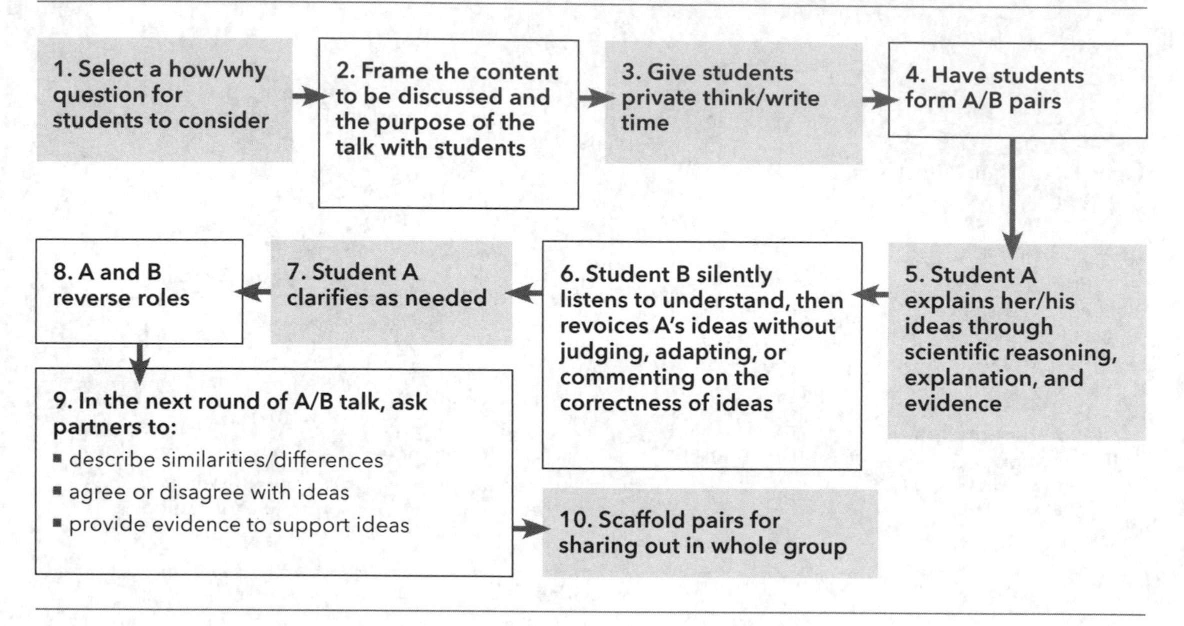

In the past, teams of teachers working on this subpractice agreed that it requires more than just asking students to do turn-and-talks. They specified four "non-negotiable" elements that had to be present in any variation of the practice:

- Talk turns are structured, often with a set amount of time, and specific roles are made explicit for students.
- Structured talk should ask students to extend beyond "what-level" explanations (just descriptions) for science events or processes.
- Each student is required to share her or his own thinking.
- Talk is open-ended and encourages students to share multiple responses.

Teachers then identify a part of the sequence and a tool they are most interested in studying in relation to student learning. Teachers in the team will then write a *reasoning stem* to describe their theory of how students learn and what supports their learning. We commonly use this stem (blank spaces are shortened for readability): "If I want to improve *students' scientific explanations*, then I need to focus on _____ and a way to do this is to _____ and/by/with [if applicable]

____." We suggest that teachers start by writing their own stem independently and then rejoin the group to collaboratively compose a final version. Teams can consult earlier chapters of this book or seek additional educational research articles when formulating these hypotheses. As teams engage in PDSA cycles, they can revisit their hypotheses with data and begin to address the key questions: Which practices work best in our classrooms? Under which conditions? And for whom?

In many ways, laying all this groundwork has already had the potential to improve teaching. Consider, for example, that colleagues have reasoned collectively about why certain learning outcomes are particularly valuable for their students, discussed what subpractices might support student reasoning and participation on the way to this goal, talked openly about how they've used these practices previously, redefined key parts of the practice together, and developed supportive tools for students. These activities reflect explicit attention to everyday routines, and deepen everyone's understanding of them.

Developing Practical Measures

Finally, teams determine how they will collect data during the enactment of practice. The data is often about what students do in response to instruction, and will be key to measuring the team's progress toward improvement goals. For the first PDSA round or cycle, teachers might invite colleagues to their classroom to record informal observations, but for subsequent rounds they will want to develop *practical measures* (meaning doable) for processes and outcomes—in other words, "measures for improvement."[4] Practical measures provide teachers with frequent, rapid feedback that can be used to assess the impact of tools and teaching, and to change them in strategic ways. The data collection methods need to be relatively undemanding and should take only about five to ten minutes.

Teams should develop two types of practical measures: indicators of processes (what we see teachers and students doing in classrooms), and indicators of learning outcomes. Process measures might be as simple as a tally of the number of times the practice (or part of a practice) or tool was used and when it was used (the beginning, middle, or end of a unit). Outcome measures should focus on what students are learning and how they are experiencing the practices. In our work, we collect data on students' depth of science reasoning, with a focus on ELLs' use of language. As part of this effort, we collect data about what particular students say during small-group work, because we have found they often talk about much more science than they end up writing. Table 13.1 shows a data collection tool we use for this, particularly in the Plan and Study phases of the PDSA cycle. We also collect samples of student work (not necessarily work sheets) that can be analyzed after the lesson.

Another convenient way to collect data is via exit tickets/slips that can be given on a weekly basis, so that students can provide feedback on how they are learning. This type of data seems particularly important during the first few months of school as teachers are setting up discussion norms in classrooms. Figure 13.4 shows an exit ticket that one PLC team used as a practical measure for how students were participating in and experiencing structured talk for "how" and "why" reasoning.

TABLE 13.1 **Practical measure for assessing depth of students' scientific explanations during small-group work**

	WHAT: Student describes what happened. Student describes, summarizes, or restates a pattern or trend in data without making a connection to any unobservable/theoretical components.	**HOW:** Student describes how or partially why something happened. Student addresses unobservable/theoretical components tangentially.	**WHY:** Student explains why something happened. Student can trace a causal story for why a phenomenon occurred or ask questions at this level. Student uses important science ideas that have unobservable/theoretical components to explain observable events.
What might this sound like today?	[Teacher fills in ahead of time]	[Filled in ahead of time]	[Filled in ahead of time]
Student 1: ☐ intermediate ELL ☐ advanced ELL ☐ not ELL			
Student 2: ☐ intermediate ELL ☐ advanced ELL ☐ not ELL			
Student 3: ☐ intermediate ELL ☐ advanced ELL ☐ not ELL			

FIGURE 13.4 **Practical measure: Exit ticket used at the high school level to understand students' experience with structured talk for how and why reasoning**

STRUCTURED TALK–Student Exit Ticket

1. When you engaged in structured talk with a partner, which of the following did you try? (check ALL that apply)
   - [x] I shared my idea
   - [x] I listened to my partner's idea
   - [x] I lagreed with my partner's idea
   - [ ] I added on to my partner's idea
   - [ ] I disagreed with my partner's idea
   - [ ] I used scientific evidence to support my idea
   - [x] I asked a clarifying question
   - [ ] I could revoice my partner's idea
   - [ ] My partner and I looked for similarities and differences in our ideas
   - [ ] I used a sentence stem to explain my idea
   - [ ] Other
2. What did you and your partner talkabout? Be specific

   we tallced about what situations are negative or positive feedback

   "We talked about what situations are negative or positive feedback"
3. What went wrong in your discussion? What could have gone better?

   we agreed on the decisions but we need to clarify

   "We agreed on the decisions but we need to clarify"
4. Explain one thing in this unit that you understand better or differently after talking with your partner today.

   I learned now others think and why

   "I learned how others think and why"

To receive rapid feedback from younger children, we suggest reducing the writing requirement and having them circle images that represent their experiences, as shown in figure 13.5.[5]

Data from these practical measures will become the object of study during monthly team meetings. With a focus on specified practices and tools, on theories of how students learn, and on data, teams are ready to engage in multiple rounds of PDSA cycles. We have found that most teams focus on *one* practice during a year, but with each PDSA cycle, they often modify tools or practical measures. Through rapid cycles of development and research, teams can fail fast, make changes, test again, and improve rapidly.

FIGURE 13.5 **Two practical measures: Exit tickets used at the second-grade level to inquire into students' experiences with various discourse practices in science class**

Circle how you felt in science class today (you can circle more than 1 feeling).

Excited	Frustrated	Happily Challenged	Confident	Creative
I couldn't wait for science to start.	I got a little mad because the work was too hard.	The work felt just right (not too hard, not too easy).	I felt like I could teach someone else.	I had good ideas in science class today.
Interested	**Confused**	**Bored**	**Scientist**	**Helpful**
I liked the learning today and want to learn more.	I didn't understand.	I couldn't wait for science to be over.	I felt like a scientist.	I helped some of my classmates with science.

Name ______________________ Date ______________

Circle the face that best describes how you felt today:

	STRONGLY DISAGREE	DISAGREE	DISAGREE A LITTLE	AGREE A LITTLE	AGREE	STRONGLY AGREE
I had a chance to figure things out on my own in science class today	☹☹☹	☹☹	☹	☺	☺☺	☺☺☺
I had a chance to decide how to do things in science class today	☹☹☹	☹☹	☹	☺	☺☺	☺☺☺

What did you learn in science class today?

____

____

STARTING THE PSDA CYCLE: THE PLAN PHASE

Teams will want to determine a monthly plan for enacting the subpractice under study and using the practical measures tools with their students. They will also want to decide how frequently to collect data on the process and on student learning. Not all team members need to adopt the same schedule. In some cases, it might make more sense for different grade-level teams to form their own approaches to trying out subpractices and data collection.

Box 13.1 is a sample planning template teachers can use during PLC time. Plans can be developed individually, but other teachers and participating coaches on the team should have a chance to co-develop or refine it. This way, more minds are at work on it and more people feel ownership. If the design goes well, everyone can celebrate the success, and if it fails, then at least it was a group effort in which everyone learned together.

Important features of this sample plan include: identifying the part of the lesson and unit where the subpractice will be used, identifying specific tools

BOX 13.1 Practice plan for structured talk for how and why reasoning

Date: ___________

Placement in lesson (circle one): Beginning, Middle, End

Placement in unit (circle one): Beginning, Middle, End

1. What is the purpose of the talk planned for this lesson?
2. What, if anything, are you changing about the talk opportunities from last time? What do you hope these changes will do?
3. What is the specific open-ended question that students will answer during the allotted structured talk time?
4. What tool(s) will students use to support their participation/learning?
5. Anticipate what you might hear in student reasoning at different levels of the explanation:
   - What:
   - How:
   - Why:
6. Will I collect data on this enactment? If so, what kind?

and questions that will be asked of students, anticipating student thinking, and determining if and how data will be collected. These criteria can be used to plan any subpractices teams choose to focus on.

As part of the Plan phase, teachers will want to design a *tool* or *suite of tools* that support the practice, such as a model template or explanation checklist, along with the supports for the teaching practice.

INTO THE CLASSROOM TOGETHER: THE DO PHASE

Now you are ready to carry the work into the classroom. Each team member enacts the lessons they planned and collects data. We have found that teachers have more success when they include the practice or subpractice goals in their unit plan, then select specific dates and class periods to enact their plan.

Team members may also decide to co-teach lessons using a *Studio Day* model. In this model, teachers get release time to meet together in the school. One member is the host for the day and will teach two lessons that will be observed by others. Importantly, we have found it helpful for coaches or lead educators to plan with the host teacher a few days prior to the lesson being observed—this spreads the risk around.

The team begins the day by reviewing the science behind the lesson; they might draw a consensus model of the focal idea together, or make a list of possible student explanations for the event or process being studied. This prepares all teachers to listen for a wider range of productive science ideas from students. Everyone then discusses the subpractice being used and the tools for supporting student reasoning. The host teacher then enacts the lesson with a class of students. During this time, other team members can cofacilitate student activity in small groups (optional), videotape student conversations (unobtrusively or they can interact with learners), and collect samples of student work. Midway through the class period, we occasionally engage in a "teacher time-out" where everyone huddles to review the questions teachers are using, discuss the ideas students have shared, and assess the plan for the remainder of the lesson. Often we adjust which questions should be asked in a whole-group conversation at the end of the lesson or what scaffolding might be used to generate more talk from a wider variety of students.

After the class we debrief the experience through discussion, and look at artifacts of student work or review video clips. In this debriefing, the facilitators (lead teachers or coaches) guide peers through conversations aimed at understanding how students were or were not learning/participating, and the implications for their own instruction. We then make adjustments and reteach the lesson to a different class later in the day, applying new insights about the task, talk, and tools used in the first lesson. Finally, teachers reflect on these changes and develop instructional plans for using what they've learned.

The Studio Day model enables faster PDSA cycles and affords important opportunities for teams to learn collaboratively. During the school year, teams may enact several Studio Days; some years we have run as few as two Studio Days, and others up to six (depending on the level of funding support and the number of teachers on the team). Our Studio Day model, then, involves: (1) teacher leaders or coaches visiting a few days ahead of time with the host teacher to co-design a lesson; (2) team members gathering on Studio Day to get an orientation to the next few hours of work; (3) teams unpacking scientific content to be taught and developing a shared understanding of different levels of a scientific explanation that students might create; (4) teams observing the classroom; (5) teams debriefing how students are learning and/or responding to the scaffolds designed for that day and making adjustments to instruction; and (6) teams observing the second class with summary team conversation to follow.

DRILLING INTO DATA: THE STUDY PHASE

For the Study phase of the PDSA cycle, teams set aside 90–120 minutes about once a month to examine data from their practical measures. In our work with teachers, we begin with an analysis of students' scientific explanations and collectively assess student work. This can be done for an entire class section or just for focal students. Figure 13.6 shows a What-How-Why chart (see appendix D) generated by fifth-grade teachers. The left side was created as part of their "gearing up" for the lesson by going over the science content together prior to their observation of a class. This Studio Day took place about midway through the unit, so students had learned a bit about forces and how to represent them with arrows, had touched on gravity and friction, and had recently learned about and applied the idea of balanced and unbalanced forces. The lesson had students revisiting the model now that they were midway through the unit. The teachers assembled the right side after their observations using collective data; each sticky note represents a student (or pair of students) that teachers observed

FIGURE 13.6 **Team analysis of students' scientific explanations from a fifth-grade lesson**

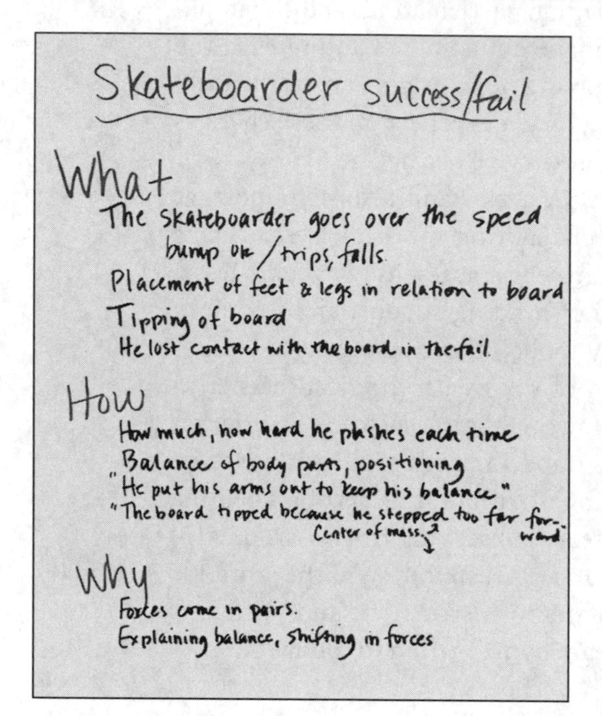

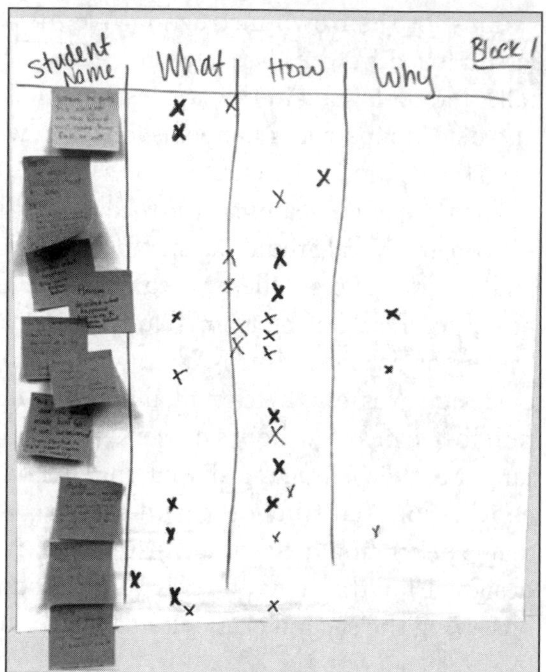

and where, based on the rubric, they thought those students were in their level of understanding.

Team members then compare and contrast student performance from one PDSA cycle to the next, and use data from other practical measures to consider why there are similarities or differences. Then the team considers the following questions:

- What did we learn from the data?
- What parts of the practice seemed to be working well, or not, for our students?
- Which students seemed to benefit?
- What did we notice about how ELL students participated in the lessons?
- What is still puzzling us about this practice?
- What might we try next time to better support student learning?

Facilitators record these questions on poster paper and team members each write their insights in a "chalk talk" format. Teachers then select one idea to share out to the team. To finish, teachers consider what they learned that is generalizable to their practice.

NEXT STEPS: THE ACT PHASE

In this phase of the PDSA cycle, team members consider: *What might we try next time to better support our students? What new questions do we have?* By "try next time," we don't mean some new random teaching experiment, but rather a data-informed next step. A team can take a look back at their driver diagram and ask, "What part of this have we made headway on?" It's rare that any team can claim mastery over a practice they feel is important or feel that a tool they've tested is now a perfect product. It is more likely that the latest round of PDSA has revealed weak spots in the practice or indicated that a tool may be well designed but is ineffective in classrooms unless used in conjunction with a particular routine. The next cycle of investigations can use these outcomes as a launching point for another round of inquiry into teaching. Team members may make a commitment to implement a new change or to stay with a practice/tool and investigate more deeply how it supports student learning.

Here is an example of action a team of seventh-grade teachers took after noticing, during a Studio Day, that students needed more support for reasoning and writing about the flow of electrons in chemical batteries. Between the first and second classroom observations of the day, teachers added scaffolds to the model template by creating circles for students to "zoom into" parts of the batteries (only a small portion of the students spontaneously included zoom-ins in the first iteration) and boxes to support reasoning at the what, how, and why levels (most students reasoned only at the what level during the initial lesson). Interestingly, they made these changes to better support a class with a high number of ELL students, and after analyzing student work, they found the ELLs outperformed the native English speakers in another classroom who did not receive these improved supports. They concluded that non-ELL students could benefit as well from these adjustments, prompting the team to generalize that some scaffolds are necessary, regardless of the student population.

As a school year progresses and investigations into practice accumulate, useful findings ideally build on one another. Teams will want to create representations of their learning to act as a reminder of what they've worked on, and to share with school leadership. We suggest a short (one-page) account of each PDSA cycle with a description of the focal subpractice and how the team's work supported school goals, department goals, and individual growth plans. We also suggest including images of student work and key tools used across classrooms. Principals will want to understand what work you've done, and how it serves aims they recognize as important.

CASE STUDY OF PDSA CYCLE IN HIGH SCHOOL: STRUCTURED TALK FOR HOW AND WHY REASONING

A high school team wanted to support more substantive student-to-student talk. They implemented the *structured talk for "how" and "why" reasoning* practice and began to use it on a weekly basis (reminder: collecting data can happen regularly and outside the context of Studio Days). At the end of some classes, they used an exit ticket to understand students' perspectives on the new ways of talking with one another. The team evaluated twelve hundred exit tickets across several classrooms and discovered an interesting pattern: few students reported disagreeing with one another. Teachers were curious about this pattern, so one of the team members decided to interview students about what they thought disagreeing sounded like. Surprisingly, students gave a range of responses, including cursing and being disrespectful to one another. This solved one mystery about why productive differences of opinion weren't explored, and gave teachers an entry point for talking with students about disagreeing in science and how it might be similar to or different from disagreeing in other everyday settings.

In the next PDSA cycle, two of the teachers decided to role-play respectful disagreement during their instruction as the basis for conversations about ideas. After their demonstration, they listened carefully to student discussions and asked a few students to perform a "replay" of their productive conversation to the entire class. Teachers continued to track this disagreement item on the team's common exit ticket. Students reporting productive disagreements with a peer jumped from 10 percent in November to 40 percent in February (see figure 13.7).

FIGURE 13.7 **Change in exit ticket data from November to February, after supports for scientific disagreements were introduced**

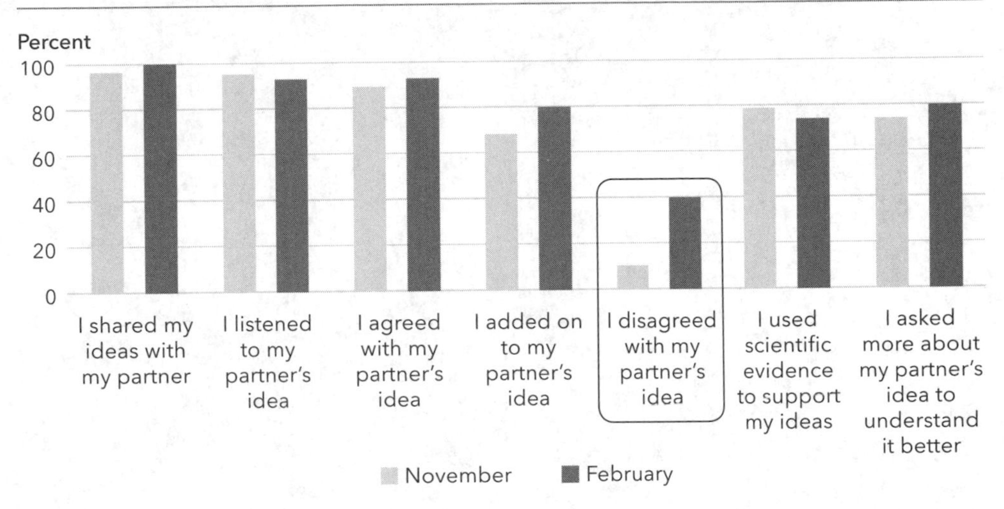

The moral of this story is: if you don't use data to find out what the problem is, you can't solve it. In this case, building students' capacity to engage in argumentation was important enough to investigate and experiment with.[6]

FOURTEEN

Can We Be Ambitious Every Day?

OVER THE COURSE of a school year, teachers and students navigate through a dozen or so units and a wide range of science topics. This raises a question: Can we teach ambitiously every day? Perhaps a more pragmatic way of putting this is: Which teaching routines should make an appearance several times a year, but not necessarily in every unit? Which of the techniques and tools we've described in these chapters are so fundamental for supporting learning, regardless of the science topic, that they should be used every week or even daily?

AST ROUTINES THAT SHOULD BE USED SEVERAL TIMES PER YEAR

Let's start with the AST routines that students should experience several times per year (see figure 14.1). Number one would be the chance to develop and revise evidence-based explanations for complex events. These activities are foundational for deep learning and for understanding science as a discipline. As we've argued previously, causal explanations are the ultimate goal of science, and when students are asked to do this work, they learn content in generative and flexible ways, come to understand how evidence is used in the discipline, and see how to revise their ideas. We recommend that you integrate ambitious science learning experiences, focused on developing gapless explanations for

FIGURE 14.1 **AST routines that should be used as the "backbone" of the curriculum, several times per year**

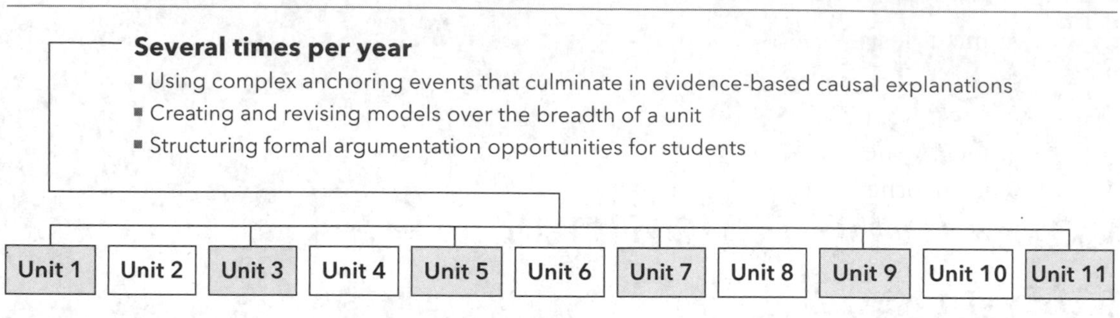

anchoring events, into the curriculum several times per year rather than every unit because (1) it is labor-intensive to develop high-quality anchoring events that align with standards and other big ideas; and (2) there are other ways to organize units for rigorous learning outcomes (discussed later).

It is more reasonable to aim for revamping one or two units each semester in your first year—not getting them to polished perfection, but just ready to pilot-test. These unit designs should include a compelling anchoring phenomenon and the lessons that will underpin students' evolving explanations. Then, in your second year, take time to improve upon those units while also working to add one or two more redesigned units per semester. This would give you four to six well-constructed units within your first two years and multiple chances for you to learn from and improve them. This pattern can be even more productive if you work with a grade-level or course partner teacher, or even colleagues from other schools who teach similar courses.

Second on our list of AST experiences you should provide several times per year is creating and revising models over the breadth of a unit. For the same reasons we mentioned earlier, however, you don't have to design the full modeling experience into every unit for students. Some science phenomena can easily be modeled, but others pose real dilemmas for teachers and students to express graphically. Even if model templates and puzzling events could be developed for every topic, students will understandably get piqued if faced with constant requests to "draw and redraw."

The third kind of experience students should have several times per year, but not necessarily in every unit, is formal instruction about argument. By "formal," we mean lessons that are built around argument as a scientific practice and designed to help students learn to make claims, express uncertainty about findings, consider alterative points of view, justify what they think they know, and respond to the critique of others. The emphasis here is on "learning to argue." On the other hand, "arguing to learn" opportunities should be incorporated into *every* unit. Students can use the language of claims and justification, for example, to deliberate about the best ways to design an investigation or represent data to an audience. Nearly any lab activity can be an occasion to use this rhetoric.

When we advocate that these powerful experiences (improving explanations for anchoring events over time, revising models, learning to argue) be interspersed regularly throughout the year, we are acknowledging that there are other ways to teach for understanding that don't follow the AST framework. Students should have opportunities to learn how ethics and social justice can be lenses for looking at science and at real-life issues that affect people. For example, entire units have been developed around the ethics case of Henrietta Lacks, an African American woman whose cervical cancer tissues were the source of *HeLa* cells—the first "immortalized" cell line and one of the most important used in medical research to the present day. Lacks had a tumor biopsied during treatment for cervical cancer at Johns Hopkins Hospital in Baltimore, Maryland, in 1951. No consent was obtained to culture her cells, nor were she or her family compensated for their use. There are "big ideas" here that students can unpack, some of which have to do with cellular biology and others that have to do with individuals' rights to determine the privacy of their medical history and the fate of the very tissues that make up their bodies. Such units make space for courageous—and critical—student conversations about social injustices related to the past, present, and future of science. Other units, focused on social justice and the environment, are becoming commonplace. Students can spend weeks going deep into questions like "What are the ecological and social impacts of allowing different types of mining in our state?" or "How might climate change affect agriculture and fishing around the world, especially in high-poverty nations?"

There are still other ways of structuring units that don't follow the typical AST trajectory. *Project-based learning* (PBL) is an example. In the PBL framework, students address science-related challenges in the community and create authentic products or solutions that are informative or useful to audiences outside of school. These units might address questions like: "How can we reduce energy consumption in our school?" or "Should we retrofit local buildings to withstand earthquakes?" In one of the schools we work with, teachers are pairing PBL curricular goals with AST practices around the question: "How is it possible for plants to grow without soil?" They are helping students develop deep understandings of a hydroponic gardening system in their school, where plants can be grown in water that is also habitat to different species of fish. Some of the subquestions focus on the biochemistry of the water as it interacts with air and sunlight. Students add their own questions, such as "Why do we need the fish?" Over the course of this unit, after students have developed models for how and why plants can grow successfully in the system, teams then focus on agriculture, aquaculture, and sustainability questions, such as: "How can we organize our hydroponic garden so we can provide a steady supply of fresh greens to our local homeless shelter's kitchen?" and "Would hydroponic farming help provide a sustainable way of growing food with less water than traditional agriculture along the Front Range where drought is commonplace?" and "Could this strategy also help provide fresh produce in 'food deserts' in urban and rural Colorado?"

Another model for teaching guides students through the historical development of foundational science ideas. These include the evolution of models of the atom (chemistry), development of theories about light (physics), plate tectonics (earth sciences), and natural selection (biology).

Finally, citizen science is becoming increasingly popular in schools. This involves collecting data for real research projects under way across the country. Students, parents, or any member of the community can observe migrating birds or measure the lead concentrations in local drinking water and upload their data to its appropriate national database. Teachers can easily form instructional units around citizen science projects that help students understand what counts as data or see what claims can be made about natural or manmade phenomena. Along the way, students can learn a great deal of science content.

All of these kinds of units—social justice-focused science, PBL, historical roots of science ideas, citizen science—represent rich opportunities to learn. And even within these types of units, you can still use core elements of AST to make them more engaging, as we describe in the following section.

AST ROUTINES THAT SHOULD BE USED ONE OR MORE TIMES EACH UNIT, REGARDLESS OF TOPIC OR LEARNING GOALS

Some AST practices should be part of every unit, no matter what the science topic is or what the larger aims of the unit are. In figure 14.2, we identify practices or routines that all have to do with supporting students' development of ideas over days and weeks, regardless of the unit framework you are using. Although formative assessment is listed as its own item, you can use nearly every routine in the "one or more times each unit" category to understand where students are in their thinking, give students feedback, and inform changes in your teaching.

Most of these routines affect learning more profoundly if they are used every few days rather than once a unit. The impact on students' reasoning and participation also builds when you use some of these routines together, such as making individual sense of activity (by moving among the tables during small-group work and pressing students for reasoning) and keeping public records of how knowledge is accumulating, changing, and being reorganized by the whole class (by adding to a consensus model, a list of current hypotheses, or simply a chart of "what we think we know now" at the front of the room). These should be part of any unit you teach.

FIGURE 14.2 **AST routines that should be used in every unit of instruction**

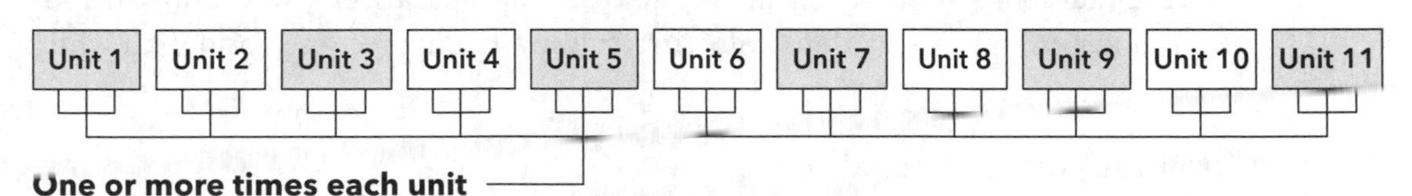

- Eliciting students' ideas, experiences, and language to use as resources
- Supporting students' sense making through small-group and whole-class discourse
- Creating opportunities for students to use ensembles of science practices
- Keeping records (public and individual) of how knowledge is changing or being reorganized
- Asking students what information or experiences they need to make progress
- Using formative assessment strategies to give students feedback and guide instructional decisions
- Helping students understand science ideas by using representations, explanations, and relevant vocabulary
- Allowing students to show what they know in multiple formats

AST ROUTINES THAT SHOULD BE USED EVERY DAY

The useful-every-day tools, practices, and routines of AST all address students' basic needs to understand why they are being asked to engage in certain activities, to feel their learning is meaningful, and to believe they are capable of participating (figure 14.3). Every student, every day, must know what they are being asked to do (framing lessons and activities helps here). They have to feel like they are learning something that builds on previous ideas and toward something on the horizon (lessons clearly contribute to larger goals beyond the day). And students all have to feel that they can participate in the work of the day (via scaffolding), whether it is joining a discussion, writing about how they might design an investigation, or collaborating with peers during a lab activity. Some of our elementary teachers have pointed out that these routines are not science-specific, but rather are useful across the subjects they teach. In their classrooms, students benefit from a consistent set of expectations, routines, and community-building efforts. Regardless of age, all students need to feel secure about their roles in the classroom. This is the work of teaching that can go beyond our practices and tools, in part because it's about maintaining trustful relationships with students. Without a sense of safety and competence, young learners will be reluctant to take the risks necessary for growth.

STARTING OFF THE YEAR

Building in all three levels of practice in figures 14.1 through 14.3 requires planning that looks at the whole academic year. In preparation, we recommend you sit down with peers, get out your curriculum (if you have one), and plan which

FIGURE 14.3 **AST routines that should be used every day**

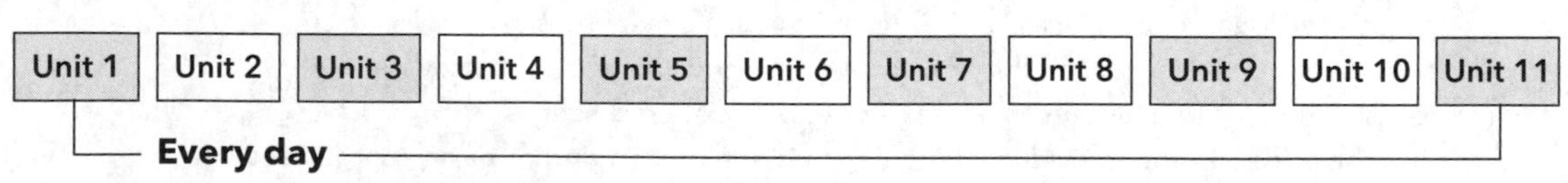

- Designing lessons to build toward bigger goals that students can comprehend
- Framing lessons so all students know why the activities are relevant and know how to participate
- Giving students chances to talk about ideas rather than only procedures or vocabulary
- Scaffolding opportunities for students to talk, write, and participate with peers

unit comes first, second, third, and so on, making these decisions based on where you can incorporate anchoring events, modeling, argumentation, and major investigations. You can space, throughout the year, opportunities for students to engage with your most puzzling and complex phenomena. You can introduce them early to modeling practices or hypothesis lists where ideas are revised over time, and give them chances in other units to become increasingly independent at this kind of work.

The first year you try this, you may have only one or two units where you can build in these challenges. That's okay. Each successive year—and with the help of colleagues—you can adopt or adapt more units of high quality that form solid foundations for the courses you are teaching. Over the year, your students will experience the tools, talk, and routines associated with knowledge building, getting better at these with each unit.

Some topics may be better taught as mini-units in which students get brief and highly focused exposure to a small set of ideas that don't fit easily into an anchoring event context. For example, in preparation for a high school ecosystem unit, one of our teachers was excited about decomposer biology; however, he could not find a natural fit for these ideas in a curriculum that focused so much on interrelationships between plants, animals, and the environment at a macro scale. His alternative was to design a five-day mini-unit that featured a dynamic ecosystem at ground level in a forest, where worms, fungi, and bacteria moved energy and biomass around in ways that made them indispensable to the health of all other organisms, large and small. In such mini-units, the routines in this chapter's figures still apply. Even though the lesson sequences are short, students can still learn in meaningful ways.

As you plan for the year, carefully design students' experiences for the first few days, because they are quick to form opinions about "What kind of work will we be doing in *this* classroom *this* year?" and "What is science about in *this* classroom?" Some of our teacher colleagues start by sharing a brief representation of the science field they'll be studying together (for example, chemistry—what exactly is it, and why should we study it?). They also identify the subfields of inquiry using student-friendly language, and then lay out typical questions that scientists in these areas pursue. They make these as relevant to students' lives as possible. By itself, these moves are eye-openers for a lot of students, who had never thought of scientists as having genuine questions or being puzzled. When we visit these classrooms, we can see the wheels turning in their heads:

"So *not knowing* is okay? Cool." Our teachers also share with students the different ways that scientists collect data in a specialized subfield and then represent their findings to the public. Again, this is an epiphany for many students. They often ask great follow-ups like "Why *aren't* scientists studying X or Y?" or "What happens if you can't collect certain kinds of data?" These conversations help make science come alive for students, and head off the view that it is a conglomeration of technical vocabulary, formulas, and irrelevant facts.

In your first year or two of doing AST work, you'll notice big differences in what your students can do over time. They'll start to work more independently with argument, the use of evidence, modeling, and the design of investigations. They'll gradually take up certain forms of talk without prompting (making claims, asking for clarification from peers, adding on to what classmates say), and they'll become more capable of supporting their peers in doing challenging work. Even more exciting is witnessing what students can do when a whole school or department uses Ambitious Science Teaching at successive grade levels. In some schools, for example, teachers find that second graders can reason deeply about evidence, explain phenomena using abstract ideas, and design investigations based on specific questions they have. Imagine if these children had opportunities for this kind of learning in every grade level. By the time they got to high school, they would require a curriculum and classroom experiences that are vastly different from what is the norm today. Science learning at these secondary levels would require a radical transformation of teaching.

This vision can become a reality if we stay committed to changes that matter in the classroom. We have to work with colleagues to further our own understandings and skills, but we also have to assert ourselves beyond the classroom to help administrators, parents, and policy makers reimagine what science learning can look like in the twenty-first century. Everyone, then, shares the responsibility to learn something new.

APPENDIX A

Coherence Between AST and Professional Standards for Practice

CHANGING WHAT YOU DO in the classroom is risky business, no matter what your level of experience. Teachers are held accountable, now more than ever, for providing students with verifiable opportunities to learn, and for opening up their practice for evaluation on a regular basis. For these reasons, we think it is important to point out how the instructional vision of AST is consistent with widely used evaluation systems for teachers. To make our case, we reference the Danielson Framework, which is used by school administrators across the country to provide feedback to their teachers; the National Board for Professional Teaching Standards, which were developed to certify the advanced knowledge and practices required of outstanding educators; and the edTPA, which is a preservice capstone assessment for novices that is used in more than thirty states.[1]

These evaluation instruments, and others like them, are consequential at different points during one's career, and each articulates what good teaching is by using a variety of rubrics for planning, instruction, assessment, and reflection. We argue here that teachers (or preservice teachers) who become familiar with AST should be able to look at the kinds of professional activity described in any of the rubrics and say, "I think I've got the tools to do this work."

Let's start with what the instruments have in common. All three emphasize that teachers should be able to identify big ideas in science, and use these to construct a rigorous and challenging curriculum for students. The Danielson

Framework suggests that an accomplished teacher "can identify important concepts of the discipline and their relationships to one another" (p. 11). Similarly, the National Boards document indicates that teachers should "possess an understanding of core ideas" as well as the "unifying concepts and principles that cut across all areas of science" (p. 28), and the edTPA suggests that teachers direct students to "construct and evaluate evidence-based explanations of phenomena that embody core science concepts" (p. 14). In addition, all these instruments specify that learners use scientific practices and information resources to make sense of these foundational ideas over time. When we analyze these rubrics, it becomes clear that teachers cannot attain any of the planning standards, in any framework, by using off-the-shelf materials without adaptation. Good teaching requires modifying the curriculum to students' needs, while maintaining challenging goals for learning—all of these assumptions are consistent with AST.

Three other themes, common to these evaluation documents, help define proficient teaching and reflect the priorities of AST. All of them foreground classroom talk as crucial to the learning process for students. One of the rubrics in the National Board Standards describes classrooms in which the teacher and students are "sharing ideas, conversing purposefully, and listening attentively." The teacher "equips students with skills that support collaboration, such as the ability to ask thoughtful questions and respond respectfully to one another's ideas" (p. 16). All the frameworks emphasize the use of formative assessment. Proficiency here means allowing students multiple ways to represent their thinking (writing, speaking, drawing, gesturing), then using this information to provide feedback to learners and make adjustments to upcoming lessons. Yet another commonality across the evaluation instruments is the emphasis on scaffolding and differentiation of instruction. References to these ideas, as central to competent teaching, are found throughout the Danielson Framework, National Boards Standards, and the edTPA.

These frameworks are consistent with one another and with the vision of exemplary teaching outlined in this book—so what value, then, do AST routines and tools add? Evaluation rubrics are good at telling us what the *indicators* are for teaching proficiency, but they are not roadmaps; they don't help us understand which pathways of practice lead to these outcomes. The Danielson Framework, for example, states that teachers' plans should "reflect recent developments in content-related pedagogy" (p. 11), but what these developments are (there are many) and how they work together to guide planning are not defined. The National Board Standards state that teachers should "promote self-directed learning and active student engagement" (p. 16), but this can be interpreted in a number of ways, and no matter the interpretation, it has to be embedded in a larger system of instruction that fosters these outcomes. The edTPA specifies that candidates provide a "challenging learning environment" with "opportunities to express varied perspectives and promote mutual respect among students" (p. 23). But again, this reads more like a goal (as it should). Knowing how to achieve this in a classroom requires that a teacher use different bodies of knowledge to draw from as well as a theory of action for creating a rigorous and civil environment for talk. AST provides this theory of action, and the necessary knowledge and tools that are grounded in research. Thus, our vision of teaching is consistent with the major evaluative standards in the profession and includes the means for achieving the highest levels of proficiency that they describe.

APPENDIX B

Reminding Ourselves of the Bigger Picture of Instruction

HERE WE PROVIDE a larger view of instruction—the "arc" of a unit. The accompanying graphics show where each of the core sets of practices is enacted during a unit, and the text describes how they work together. The overarching goals for every AST unit you teach are for students to experience learning as *meaningful* (making sense of ideas rather than just remembering or reproducing them), *cumulative* (recognizing that learning challenges each day require them to use what was learned in previous lessons and to build upon these ideas with new experiences), and *progressive* (aiming to improve explanations over time by iteratively assessing them, elaborating on them, and holding them up to critique and evidence). It might seem that well-sequenced lessons would do all this work for you, but that's not the case. You have to purposefully build in opportunities to learn in these ways, both at the lesson level and at the unit level.

Figures B.1 and B.2 represent the flow of lessons in two units, the former about wolves being reintroduced to Yellowstone National Park (tenth-grade biology) and the latter about sound energy (fifth-grade physical science). The diagrams look similar because the learning principles underlying the design of these units and the teaching practices employed are the same. Both begin with teachers identifying big science ideas from their curriculum and standards, then selecting an anchoring event, and then sequencing lessons (these activities make up core practice set 1: planning for engagement with big ideas). The

FIGURE B.1 How core practices (bottom) are used to support lessons across a unit trajectory

High school unit on ecosystem dynamics in Yellowstone. Some lessons were two class periods rather than one.

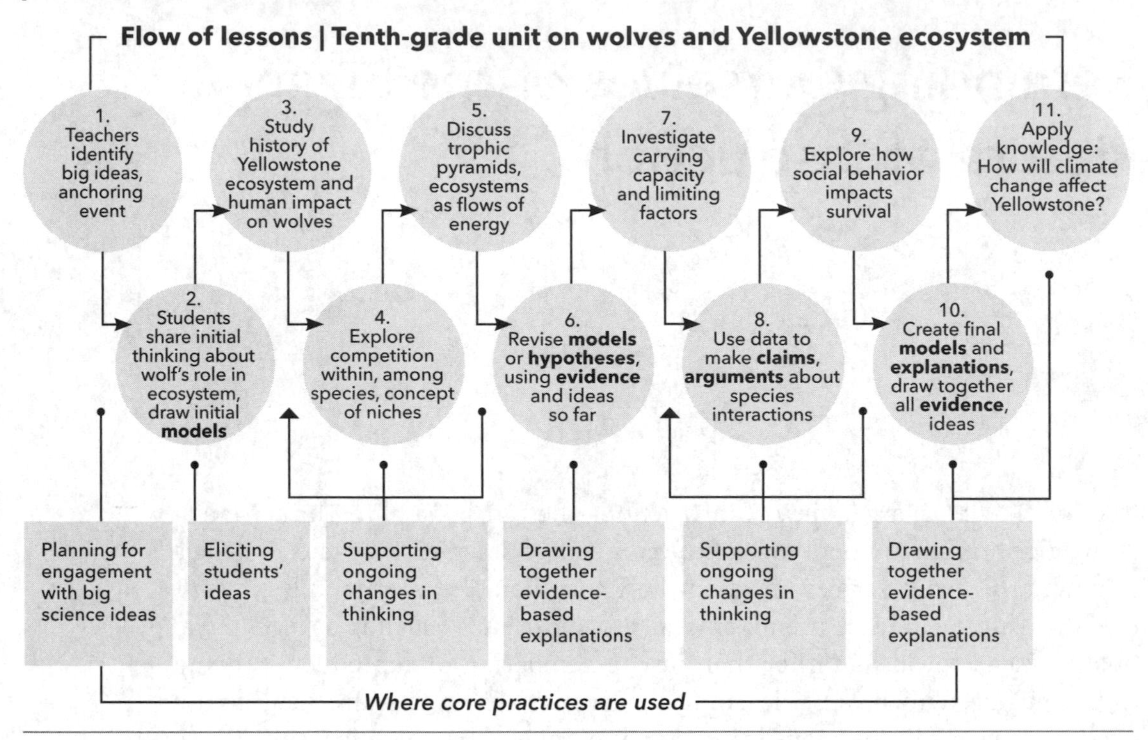

teacher starts instruction by listening to students' theories and experiences that they think may be relevant to the anchoring event. Students make their thinking public by creating and discussing initial models, developing a list of hypotheses for the anchoring event, or creating any other representations that could be updated as they learn more. The teacher uses this information to inform and modify upcoming lessons (these activities make up core practice set 2: eliciting students' ideas).

Students then embark on a series of lessons, each with its own focus (core practice set 3: supporting ongoing changes in thinking). During these lessons, carefully selected science concepts are introduced, and students engage in scientific practices for reasons the teacher helps them recognize or that they

FIGURE B.2 How core practices (bottom) are used to support lessons across a unit trajectory

Upper elementary unit on sound energy. Some lessons were two class periods rather than one.

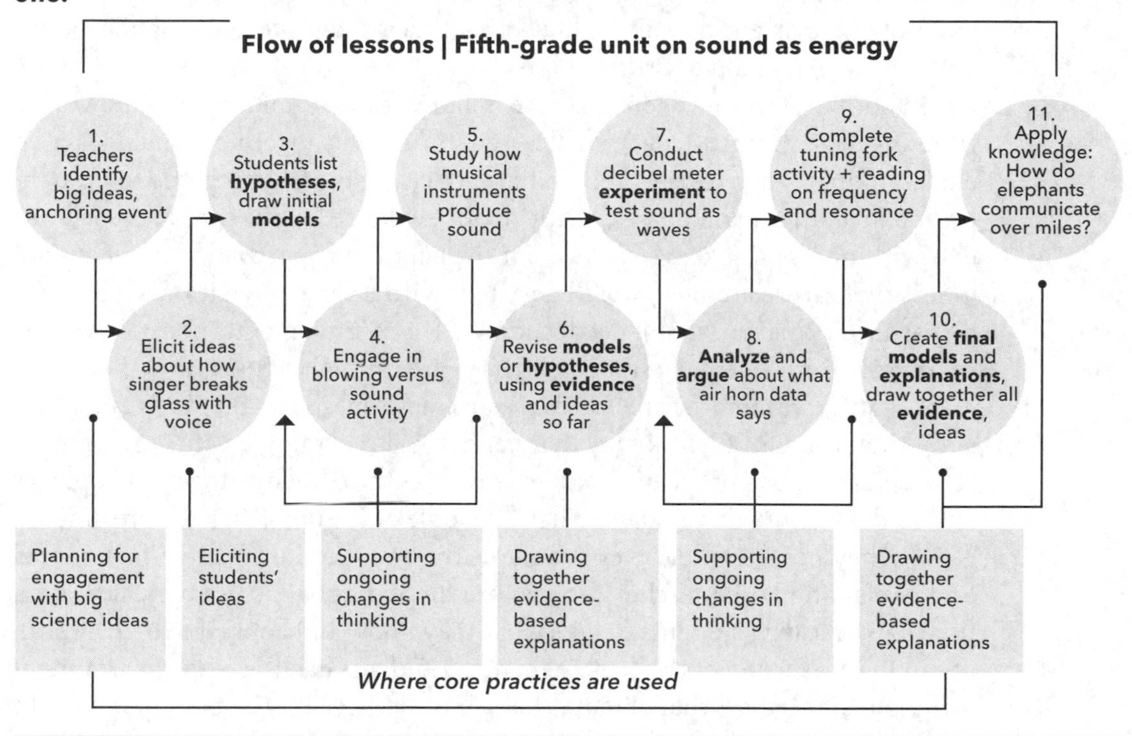

choose themselves (please don't confuse *science* practices with *teaching* practices). Sense-making discussion is supported in a variety of ways as students publicly compare new ideas against their existing knowledge and experiences. Students' theories and language can be used to modify the lesson sequence. In one iteration of the Yellowstone unit, for example, a group of tenth graders hypothesized that the wolves affected not only the death rates of prey in the park but their birth rates as well. The students argued convincingly that this should be part of the explanation; in response, the teacher dedicated a new lesson to the reproduction rates of these prey species. In the sound unit, one group of students introduced the theory that different structures and materials could absorb sound. This issue was important to them because many of their parents regularly complained to the city about nearby freeway noise and what should be done about it. In response, the teacher asked her students to go online to research noise abatement, and they then integrated what they found into an engineering conversation about how other cities had protected neighborhoods from the constant din of traffic.

In both units, somewhere near the middle, students take stock of how their thinking has changed (core practice set 4: drawing together evidence-based explanations). They revisit their initial models, lists of hypotheses, or rough-draft explanations and talk explicitly about evidence for change. These lessons end with questions like: How have our explanations improved, and what convinces us that we are now more consistent with science? What new questions do we have? What do we still need to know? These fundamental questions are as productive for elementary students as they are for high schoolers.

Fortified with more coherent ideas and new questions, students engage in additional rounds of learning activity and sense making (back to core practice set 3). Near the end of the unit, students reprise the "drawing together" of ideas and evidence work they had done in the middle of the unit (back to core practice set 4). They now assemble every resource they have to formulate a final model and explanation. The teacher can require an additional assessment task on a final day by asking students to take what they know and apply it to a novel situation. In one version of the ecosystem unit, students had to create an argument for one major impact that climate change would have on the park and why. In the sound unit, students were asked to model how elephants can communicate over many miles, using sound.

We describe these two units to show how the teaching practices build from one another, how often they are used, and what their role is in the bigger picture of learning. However, the representations we've chosen can appear a bit too tidy; for example, the outlay of the two units is almost identical. We know that real teaching, when mapped out, does not happen like this. Units are different lengths, and sometimes you have to reteach lessons or cut others short. All this is part of authentic professional work, and we acknowledge that "stuff happens" that can derail the best-laid plans for units. Our takeaway, however, is that this general pattern of teaching practices is helpful, even if it can only be roughly approximated in most units. It is, of course, crucial that you plan with big ideas in mind (core practice set 1). Eliciting is really a must on the first day (core practice set 2), although various forms of eliciting can be done on a small scale for specific ideas throughout the unit. Most of the unit is dedicated to learning activities, mini-presentations of information, the use of scientific practices, and a constant push for sense making by students (core practice set 3). In the middle, students pause and consolidate what they know, and then do the same, with some variation, at the end of the unit (core practice set 4).

We want to stress again that we are not advocating some new formula for teaching that replaces old formulas. The structure you see in figures B.1 and B.2 is a launching point for innovation. Neither the teaching practices nor the trajectory of the unit has to adhere to a set of inflexible rules. We are arguing that this framework is the most advantageous place for us all to start, and that if you want to develop new expertise, then these four sets of teaching practices, especially in combination with one another, will give you the best return on investment in terms of student learning. If you can develop some mastery over these core practices, you are more likely to create significant learning opportunities for your students. When you are ready to innovate, keep the seven elements of rigorous and equitable teaching in mind (chapter 1). Whatever lesson, unit, or tool you are designing, ask yourself: Am I anchoring students' learning in complex phenomena? Am I using students' ideas and experiences as resources for everyone's learning? Are my students engaging in science practices for reasons they understand? Am I providing regular opportunities for students to reason through talk? Scaffolding their efforts at talk, writing, and participation? Making thinking visible and subject to critique by the classroom community? Are the various learning experiences recognized by students as building toward their cumulative understandings of big science ideas? It is within these principled constraints that your innovations will be most productive and not simply different from the status quo.

APPENDIX C

Taxonomy of Tools

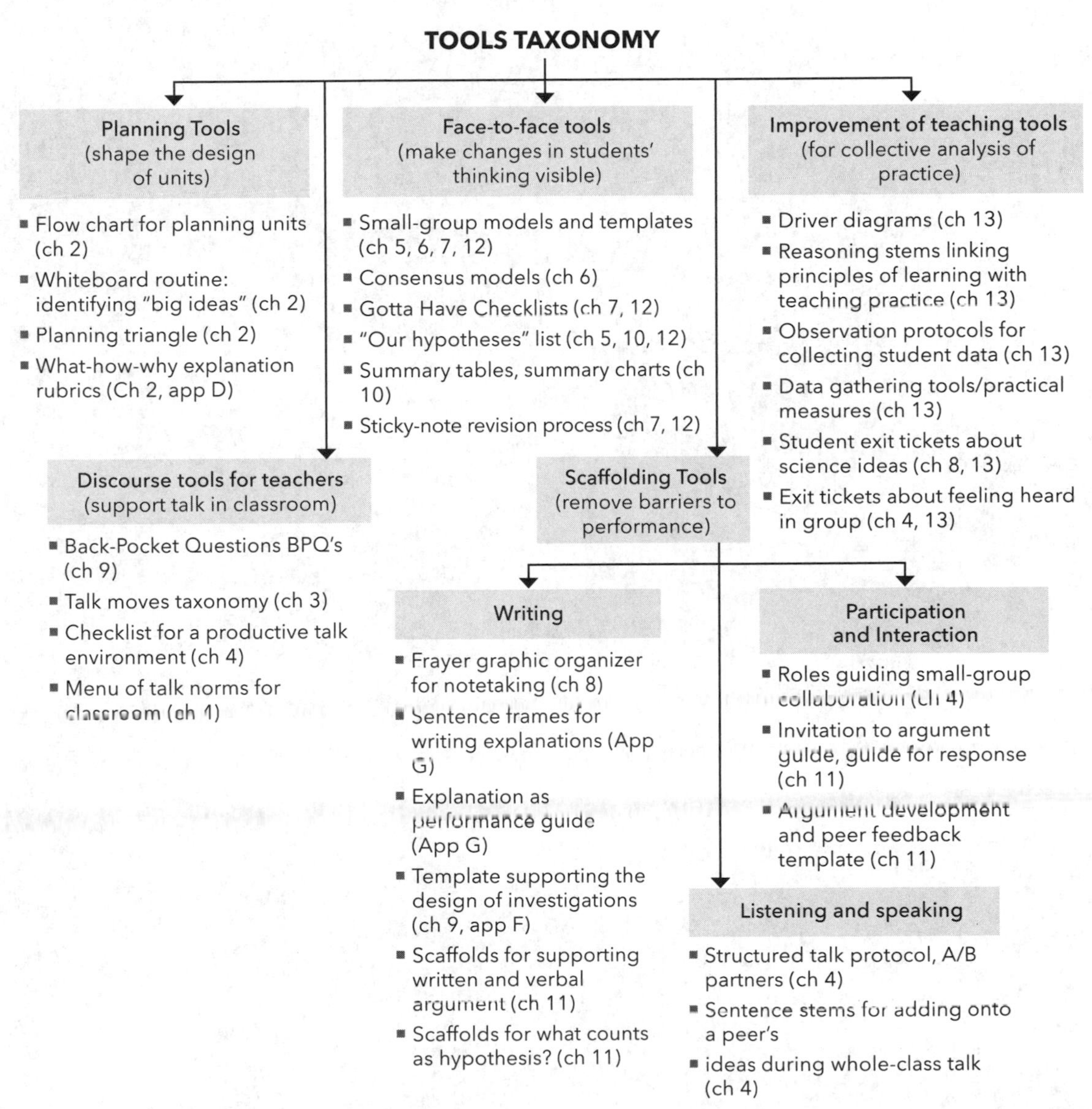

APPENDIX D

How to Help Students Understand the "What-How-Why" Levels of Explanation

STUDENTS SOMETIMES VIEW explanations as simply "right" or "wrong." But a more productive way to assess them is determining if they tell a "what," "how," or "why" story. These three levels indicate the depth of explanation, and if students can understand the differences, they can aim to write more coherent and complete causal narratives.

- *What level.* Student describes an observation or summarizes and restates a pattern or trend in data, but does not make connections to any unobservable event, process, or structure.
- *How level.* Student describes the observable conditions under which something happens *or* states a "partial why" by naming a cause and an effect without articulating what the connecting mechanisms are.
- *Why level.* Student traces a causal story for a phenomenon, using unobservable processes, events, or structures and a chain of logical steps to explain observable events.

Teachers can ask for an explanation at any of these three levels, but we consider the "what" to be more of a description. At this level, students write about observations of an event, process, or thing. Students can use their senses (sight, hearing, etc.) or use instruments (measurements of temperature, speed, dissolved oxygen, etc.) to state qualitative observations or to report quantities or counts of something. "What" stories can be elaborately detailed accounts of science ideas like plant succession (a process), cell structure (a thing), or volcanic eruption (an event), but in the end these are really descriptions rather than explanations.

A what explanation can even characterize differences between experimental groups, trends over time, or qualitative observations. Saying, "Explain what you see in the data" (and then getting patterns and trends talk) is one way to ask for this. Most teachers stop at this point, and most standardized tests set this as the limit for what students should be able to compose in terms of explanation. There's nothing at all wrong with a beginning explanation at this level; however, it only scratches the surface of what students are able to do.

In a "how" explanation, students write about relationships among variables or observations (as in a what explanation), and how these predict the ways some natural system will behave. In other words, students characterize the observable conditions under which some event or process might occur. For example, they might note that a flashlight bulb burns brighter if it is linked to two batteries as opposed to one (without mentioning energy transfers or transformations), or that a solar eclipse occurs because the moon passes between the earth and sun (without mentioning gravity, the nature of light, the ecliptic plane, or regularity of the orbits). At this level of explanation students are no longer explaining "things," but rather phenomena. Another type of how explanation is referencing unobservable events, processes, or structures that are somehow linked to what is observable in the world, but without saying what the connecting mechanisms are—for example, "Gravity causes the tides" or "DNA helps produce specific proteins in the human body." We often refer to this as a "partial why" explanation. A partial why usually involves students using vocabulary terms they've become familiar with to provide a shortcut explanation. Nonetheless, expressing these relationships between the observable and the unobservable, however "gappy," can be stepping-stones to a full causal story.

At the "why" level, students use unobservable events, processes, and entities to construct a causal narrative for why something happened. We use the term *narrative* because these explanations contain a chain of logical steps or interactions to explain observable events. The gapless explanations valued in AST are often two or three paragraphs in length.

We lay out in more detail these three levels of explanation in the context of an activity done in high school biology classes. To help students understand the concept of cellular respiration, and the fact that it happens in all cells of an organism, teachers have them work with a compound called Bromothymol Blue (BTB). This chemical, when in solution, is blue but turns clear in the presence of carbon dioxide. In the investigation, students breathe with a straw into a BTB solution before and after exercising, as a direct indicator of carbon dioxide output and an indirect measure of glucose being converted to energy with carbon dioxide as a by-product. Table D.1 provides an example of the what-how-why explanation framework that a group of teachers we worked with developed for this cellular respiration investigation. Students were asked after the lab to "explain why you saw an increase in respiration after exercise."

Note that the why level is the causal story—it requires that students theorize beyond just the BTB activity. They used their experimental results to try to piece together a comprehensive and meaningful explanation. Some teachers might look at the why level of explanation and say, "It's just got more detail," but there is a lot more than ornamental flourishes or additional vocabulary here. Every

Table D.1 Examples of what, how, and why levels of explanation

	WHAT LEVEL	HOW LEVEL	WHY LEVEL
Depth of explanation	Student describes an observation or summarizes and restates a pattern or trend in data, without making a connection to any unobservable components.	Student describes the observable conditions under which an event or process would happen or states a "partial why" by naming a cause and an effect without saying what the connecting mechanisms are.	Student can trace a causal story for why a phenomenon occurred. Student uses unobservable processes or structures and a chain of logical steps to explain observable events.
Example explanation	Our breathing increased when we started exercising by 30%, and the BTB changed from blue to yellow over the five-minute period.	The BTB changed from blue to yellow after the exercise because the body exhaled more carbon dioxide than when it was stationary. When exercising, the body requires more oxygen. As oxygen intake increases, so does carbon dioxide output.	When exercising, the body requires more oxygen, which is taken from the lungs to muscle cells (via the circulatory system and diffusion). The cells use the oxygen to break down glucose into energy and carbon dioxide. Muscles use the energy to do work. The carbon dioxide is a waste product. It diffuses into the blood and then the lungs and is exhaled. Cellular respiration happens at a faster rate when a person is exercising, so more carbon dioxide is produced. The exhaled carbon dioxide reacts with water to produce carbonic acid, causing the BTB indicator to change color.

segment of the explanation expresses a key relationship. If students can express this level of explanation in their own terms (not reproducing the teacher's or the textbook's explanation verbatim), they now understand the evolutionary purpose of cellular respiration, the mechanisms of how it happens at the cellular level, and why it is affected by internal or external conditions. Each segment of the explanation is a part of a big idea, and each is connected to the other—this is deep understanding, and more than details. The what and how explanations, while helpful platforms to get to a causal explanation, do not by themselves signal this depth of understanding.

Are students capable of this level of explanation? Absolutely, if you prepare them for it. This is why it can be helpful for teachers to take the topic of any unit and create separate what-how-why explanations that go along with some central event or process in the unit. The teachers we've worked with have filled in what-how-why rubrics for themselves, doing the most challenging why part first, then going back to fill in the what, and finally answering the how (this is the easiest way to sequence it). When it's completed, they realize "Oh! So that's what I'm aiming for. This tells me I'm going to have to give my students different readings and experiences than I anticipated, and certainly I'll have to ask them different questions along the way."

During a unit you are gradually pressing for deeper explanations. We call it *pressing* because you as the teacher are nudging students out of their comfort zone. They may be so used to playing the game of school and giving you brief, surface-level responses to your questions that they feel discomfort or unease at having to think harder. There is always pushback when you ask during a lab activity: "But why do you think you got those data? I hear you talking about the patterns and trends in your observations, but what's going on that we can't directly observe?" Students will acclimate to your strategic demands, and although the discomfort on their part never entirely goes away, their thinking and conversations will adapt to more rigorous expectations.

One of our teachers gave a what-how-why rubric (just the criteria, the top row in our figure) to his middle school students and explained the different levels. Over the next few weeks, he found his students pulling out their rubric and arguing with one another during lab activities: "You think you have a how explanation, but it's only a what!" This tells us that students can internalize differences in the quality of explanations, even to the point of using these distinctions to interact with peers about boosting their own explanations for understanding.

APPENDIX E

Rapid Survey of Student Thinking (RSST) Tool

RAPID SURVEY OF STUDENT THINKING (RSST)

Directions: Complete the RSST right after a class.

CATEGORIES	TRENDS IN STUDENT UNDERSTANDINGS, LANGUAGE, EXPERIENCES	INSTRUCTIONAL DECISIONS BASED ON TRENDS OF STUDENT UNDERSTANDING
Partial understandings What facets/fragments of understanding do students already have?	List partial understandings: __________ __________ __________ __________ __________ What approximate percent of your students have these partial understandings? __________ __________ __________ __________ __________	★ Star the ideas on list at left that need action. Instructional options: Do further eliciting of initial hypotheses to clarify your understanding of students' partial understandings Do ten-minute whole-class conversation of two or three key points elicited Write multiple hypotheses on board and/or develop an initial consensus model Other

continued

CATEGORIES	TRENDS IN STUDENT UNDERSTANDINGS, LANGUAGE, EXPERIENCES	INSTRUCTIONAL DECISIONS BASED ON TRENDS OF STUDENT UNDERSTANDING
Alternative understandings What ideas do students have that may be inconsistent with the scientific explanation?	List alternative understandings: __________ __________ __________ __________ What, if any, experiences or knowledge bases are they using to justify these explanations? __________ __________ __________ __________	★ Star the ideas on list at left that you *really* need to pay attention to, based on the following criteria: (1) Which alternative understandings seem deeply rooted (kids seem sure about)? (2) What % of kids think this? (3) Which are directly related to final explanation (not just a "side story")? Instructional options: ■ Do further eliciting about what experiences/frames of reference students are drawing on ■ Pose "what if" scenario to create conceptual conflict about validity of alternative ideas ■ Challenge students to think further/give them a piece of evidence to reason with
Everyday language What terms did you hear students use that you can connect to academic language in upcoming lessons?	Cite examples: __________ __________ __________ __________ What approximate % of your students use these terms and phrases? __________ __________ __________ __________	★ Star the ideas on list at left that you can leverage in nontrivial ways. Instructional options: ■ Use this language to reframe your essential question in students' terms ■ Use as label in initial models that you make public; work in academic versions of these words into public models and discussions later, when the need arises ■ Other
Experiences students have had that you can leverage What familiar experiences did students describe during the elicitation activity?	What was the most common everyday or familiar experience that kids related to the essential question or task? __________ __________ __________ __________ What were the less common experiences students cited? __________ __________ __________ __________	★ Star the ideas on list at left that you can leverage in nontrivial ways. Instructional options ■ Rewrite the essential question to be about this experience ■ Make their prior experiences a central part of the next set of classroom activities ■ If kids cannot connect science idea to familiar experiences they've had, then provide a shared experience all kids can relate to (through lab, video, etc.) ■ Other

APPENDIX F

Supports for Students in Making Sense of Experimental Design and Purpose

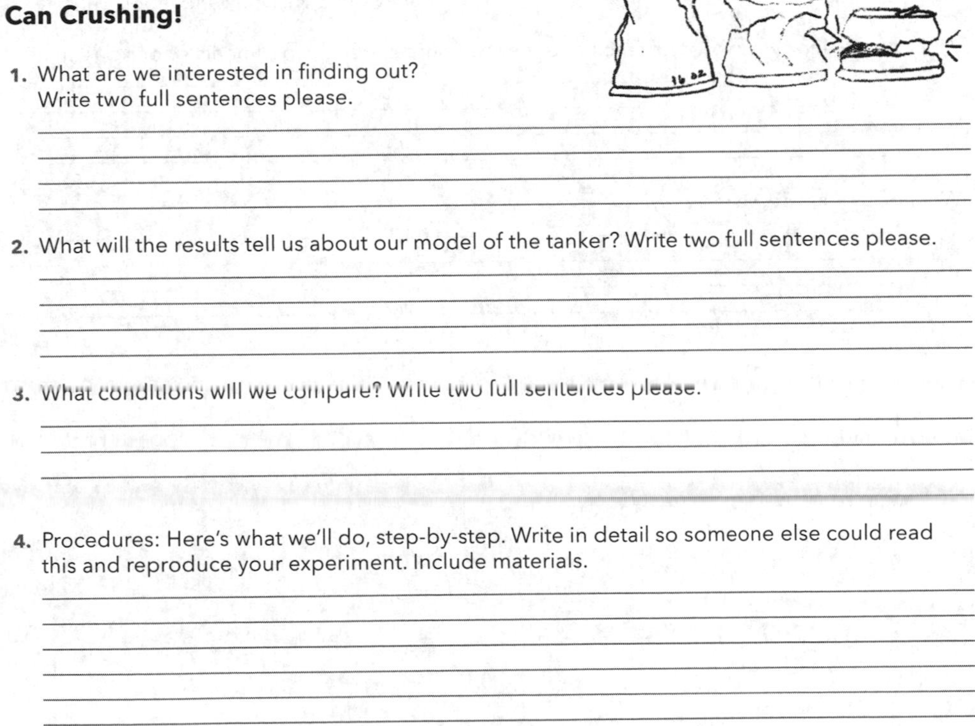

Can Crushing!

1. What are we interested in finding out?
   Write two full sentences please.

2. What will the results tell us about our model of the tanker? Write two full sentences please.

3. What conditions will we compare? Write two full sentences please.

4. Procedures: Here's what we'll do, step-by-step. Write in detail so someone else could read this and reproduce your experiment. Include materials.

Data table

What we think happened that we could not directly observe. Draw and label.

What do we think we know as a result of our experiment and the experiments of our classmates?

____

____

____

____

What questions are we left with?

____

____

____

____

APPENDIX G

Supporting Explanation Writing

HERE'S A SCENARIO you've likely experienced. You are circulating around the room as students are working on a lab activity, and you stop at a table to ask the group, "What do you think is happening here?" With a little encouragement, the students start to offer mini-theories, and in a few moments the beginning of a causal story (for the phenomenon they are working on) takes shape. Excitedly you say, "Okay, I really like your thinking; can you just write down those ideas in an explanation?" A few minutes later, when you look back at this group of students, you see their pens hovering motionless above their notebooks—they seem frozen, unable to translate their rush of ideas into written form. By the end of class they've scratched out only a few words, maybe a sentence or two, that stand in for all the promising talk you heard from them. What happened?

Explanatory writing is hard for students in part because we rarely ask them to do it at length, they rarely receive support for getting the work started, and they don't get meaningful feedback. Some of us offer rubrics for explanation writing to guide students or we show them examples of previous students' work. This is helpful, but not helpful enough.

Our approach is a bit different. Rather than focusing on explanations as an *end product*, we think of writing as a *performance*. As with any performance in sports, the arts, or business, people prepare for it strategically; they rehearse instead of simply trying it out for the first time in front of an audience. The supports you provide begin five to seven days before students are expected to "perform" for others. No strategy works like magic, but combinations of the supports we describe here seem to work best to influence writing. Over time you will find new and better ways to support learning for every student. The steps we outline are scaffolds, meaning that, over the course of the year, you would withdraw much of this help as students become more capable and independent writers. Timeframes for all of these steps will condense as students become more proficient and more comfortable over the year.

STEP 1: PROVIDE STUDENTS WITH STRUCTURED REHEARSAL TIME TO PRETHINK AND PREWRITE

The options we describe here usually last for two to three days, about a week before you expect students to have a more polished version of their explanations to share publicly. When you have students with special needs or English language learners, we recommend talking first with education specialists in your school about more specific routines that you can try in your classroom. You can also confer with colleagues in other subject matter areas to find out how they support writing. Don't do this alone.

Structured Prethinking and Prewriting

Students need *lots* of structure, scaffolding, and time to think about their ideas and to write/draw their ideas in small pieces. One strategy is to prompt students to write/draw an explanation draft in narrative form, like a storyboard. Here, students can tell/show their ideas about the beginning, middle, and end of the phenomenon. In table G.1, for example, students can sketch what they think is happening into the top row and put rough-draft text into the row below. Parsing the explanation out like this helps them focus on moments in time rather than attacking the whole phenomenon at once. It also helps you to give more targeted feedback, one part at a time.

Writing Scaffolds

Use varied scaffolds to support students' writing in the bottom row of their storyboard. We provide a mixture of generic scaffolds that students can use in any explanation for any science idea, and more specific scaffolds that are useful only in the context of a particular assignment. Here are some examples from

TABLE G.1 **Table for drafting out explanations in sequence**

BEGINNING:	MIDDLE:	END:
Sketch	Sketch	Sketch
Write	Write	Write

a middle school unit on force and motion, in which the anchoring event was about a jet taking off from an aircraft carrier:

- *Sentence starters*. These help students by giving them a "mini-structure" for constructing sentences. Examples: "The jet starts its takeoff by _____"; "One reason this happens is because _____"; "This makes me think that _____."
- *Science idea banks*. These help students by prompting for specific science ideas needed to construct a particular explanation. Examples: "In your explanation, be sure to use these ideas: push, drag, normal force, friction force, gravity." You can also use prompts in the form of items on an explanation checklist. In this case, you'd phrase a prompt something like this: "In your explanation, be sure to include balanced and unbalanced forces, and the role they play in making the jet move."
- *Cause-and-effect phrases*. These help students by providing connecting words/phrases characteristic of academic writing and expressing causality. Examples include: "This happens because _____"; "This led to _____"; "This influences _____"; "As a result of this _____." Students also need help structuring transitions between ideas; examples here include: "in addition," "similarly," "therefore," "but another possibility is," "because," and "for example." Sometimes these are given on strips of paper and student physically manipulate them, trying out different combinations before committing to writing.
- *Science concept cards*. These help students integrate science terminology into their writing. For example, during the unit, students build a collection of cards with terms, drawings, and student-generated definitions. Students are free to use these cards whenever they are working on tasks in class or on homework assignments. A "word wall" or "idea wall" can serve a similar function.

Strategic Partnerships

This prethinking/prewriting task can be done individually or in pairs where the partners have been strategically selected to support the needs of certain students. Your goal when planning these partnerships is to distribute the cognitive load across two people without totally removing the responsibility for thinking from either student.

For example, consider pairing two same-language ELLs where one student has developed slightly more advanced English skills, so that partners can switch back and forth between languages. Consider allowing ELL students to communicate in a language other than English and work on translating their text into English later as a next step. However, we would not want students to be excluded from using English or excluded from hearing others use academic English (students need to hear and use academic English in order to learn it). Use these pairings sparingly—and gradually withdraw this support as students progress in their language skills.

As another example, consider pairing certain special needs students with patient and helpful peers who could serve as sounding boards, act as writing tutors, or simply remind them to stay focused. There are, however, two important issues to be aware of. First, students won't learn to develop literacy skills if they are never allowed to write, so it is not advisable to have a student serve as a scribe for a peer with learning disabilities, unless an IEP specifically directs this kind of assistance. Second, it is beneficial to have a student serve as scribe for a peer with a physical disability, such as cerebral palsy, that makes the physical act of writing or drawing difficult.

It takes time to build a culture where students can learn to be good partners. As students develop their writing skills throughout the year, this scaffolding can be reduced and eventually removed.

STEP 2: HAVE STUDENTS ADD EVIDENCE WITHOUT ADDING MORE WRITING

Once you have rough-draft explanations or models from students, you can ask them about evidence. For example, you can point to one part of their explanation and ask, "How do you know that this part works this way?" See if students can connect the "parts" of their story back to specific lab experiences or readings. Students can add evidence by placing on their storyboard a sticky note with a short phrase about how an idea or lab experience supports their explanation. Sticky notes work well for this because they can be rewritten or put in different places.

As you work with students who have struggled in school, pay attention to your tone of voice and other social cues when asking a question like "How do you know?" Students can interpret this as a challenge or as a put-down, which may cause them to become defensive or to withdraw. Instead, explain that you want them to figure out *how* they learned about parts of their explanation and then pose your question.

This practice takes time, but it is an investment that pays off in the long run for students who find writing difficult. Plan on the "evidence adding" activities to take at least one class period. As with other scaffolds we've described in this appendix, students will need less of this support over the course of the year.

STEP 3: GIVE STUDENTS MORE REHEARSAL AND TIME TO RETHINK AND REWRITE

After students have worked out their ideas independently, they should be ready to communicate with a partner or very small group. It's probably not safe, however, to have vulnerable students trying to present in front of the whole class yet. Pair sharing or small-group sharing time allows students to do three things: rehearse their own ideas and language with a small audience; acquire ideas and language from their peers; and get feedback from you about the clarity of their explanation and their use of science ideas. You should build in some time for students to add, delete, or change ideas on their storyboards after hearing from peers and from you. Plan for this to take at least one class period.

STEP 4: THE BIG PERFORMANCE—ASK STUDENTS TO TALK IN FRONT OF THE WHOLE CLASS

If you have done the prep work described previously, students will have organized and rehearsed their ideas by now, gathered their thoughts, and tried out some academic language. When they are now asked to say things out loud (to you or to the whole class), it's not about trying to think on their feet or speak off of the top of their head; rather, it's about telling a story that has been rehearsed over a couple of days.

STEP 5: THE FINAL PRODUCT—HAVE STUDENTS CONSTRUCT A POLISHED VERSION OF AN EVIDENCE-BASED EXPLANATION

After all of this, you could ask students to revise their efforts from the previous week and write/draw a final draft of their evidence-based explanations. We often work with colleagues from literacy/language classes to develop sentence-writing, paragraph-writing, and essay-writing support that is consistent across students' subject areas. Many schools use a specific writing model to assist students with writing full paragraphs and essays.

Remember, all the strategies we've described work best in combination with one another. And exercise patience—don't expect perfection, just look for students' abilities to take some partial understandings to the next level and to develop further their use of ideas and language. It's a performance, but still a work in progress.

Notes

Chapter 1

1. Research studies are designed to investigate narrow slices of classroom life. They may focus on the kinds of student talk that happen during small-group work, the effects formative assessment has on teacher planning, or the ways in which different textbook representations of science ideas support student learning. Researchers then use data to make claims about variables in the classroom environment and how these might be related, but we have to take these conclusions with a grain of salt. In classrooms, many layers of social and intellectual activity flow together to create the human experiences of learning, engagement, or distraction. There is much more going on in our science classrooms than what can be measured by any instrument.
2. John D. Bransford, Ann L. Brown, and Rodney R. Cocking, *How People Learn: Brain, Mind, Experience, and School* (Washington, DC: National Academies Press, 2000); Ann L. Brown and Joseph C. Campione, "Psychological Theory and the Design of Innovative Learning Environments: On Procedures, Principles, and Systems," in *Innovations in Learning: New Environments for Education*, ed. Leona Schauble and Robert Glaser (Mahwah, NJ: Lawrence Erlbaum Associates, 1996), 289–325; Randi A. Engle and Faith R. Conant, "Guiding Principles for Fostering Productive Disciplinary Engagement: Explaining an Emergent Argument in a Community of Learners Classroom," *Cognition and Instruction* 20, no. 4 (2002): 399–403, doi:10.1207/S1532690XCI2004_1; Barbara Y. White and John R. Frederiksen, "Inquiry, Modeling, and Metacognition: Making Science Accessible to All Students," *Cognition and Instruction* 16, no. 1 (1998): 3–118, doi:10.1207/s1532690xci1601_2.
3. Angela Calabrese Barton and Edna Tan, "Funds of Knowledge, Discourses and Hybrid Space," *Journal of Research in Science Teaching* 46, no. 1 (2009): 50–73, doi:10.1002/tea.20269; Andrea A. diSessa, "Towards an Epistemology of Physics," *Cognition and Instruction* 10, no. 2/3 (1993): 105–225, doi:10.1080/07370008.1985.9649008; Norma González and Luis Moll, "Cruzando el Puente: Building Bridges to Funds of Knowledge," *Educational Policy* 16, no. 4 (2002): 623–41; Ann S. Rosebery et al., "'The Coat Traps All Your Body Heat': Heterogeneity as Fundamental to Learning," *Journal of the Learning Sciences* 19, no. 3 (2010): 322–57, doi:10.1080/10508406.2010.491752; Maria Varelas et al., "Drama

Activities as Ideational Resources for Primary-Grade Children in Urban Science Classrooms," *Journal of Research in Science Teaching* 47, no. 3 (2010): 302–25.

4. Hamin Baek et al., "Engaging Elementary Students in Scientific Modeling: The MoDeLS Fifth-Grade Approach and Findings," in *Models and Modeling: Cognitive Tools for Scientific Enquiry*, ed. Myint Swe Khine and Issa M. Saleh (New York: Springer), 195–218; Leema Berland et al., "Epistemologies in Practice: Making Scientific Practices Meaningful for Students," *Journal of Research in Science Teaching* 53, no. 7 (2016): 1082–1112; Richard Lehrer and Leona Schauble, "Developing Modeling and Argument in the Elementary Grades," in *Understanding Mathematics and Science Matters*, ed. Thomas Romberg, Thomas Carpenter, and Fae Dremock (Mahwah, NJ: Lawrence Erlbaum Associates), 29–53; Cynthia Passmore and Jim Stewart, "A Modeling Approach to Teaching Evolutionary Biology in High Schools," *Journal of Research in Science Teaching* 39, no. 3 (2002): 185–204; Barbara Y. White and John R. Frederiksen, "Inquiry, Modeling, and Metacognition: Making Science Accessible to All Students," *Cognition and Instruction* 16, no. 1 (1998): 3–118, doi:10.1207/s1532690xci1601_2.
5. Cindy Ballenger, *Puzzling Moments, Teachable Moments* (New York: Teachers College Press, 2009); Suzanne Chapin and Catherine O'Connor, "Project Challenge: Identifying and Developing Talent in Mathematics Within Low-Income Urban Schools," *Boston University School of Education Research Report* no. 1 (2004): 1–6; Neil Mercer, *Words and Minds: How We Use Language to Think Together* (Abingdon, UK: Routledge, 2000); Sarah Michaels, Catherine O'Connor, and Lauren B. Resnick, "Deliberative Discourse Idealized and Realized: Accountable Talk in the Classroom and in Civic Life," *Studies in Philosophy and Education* 27, no. 4 (2008): 283–97, doi:10.1007/s11217-007-9071-1; Rupert Wegerif, Neil Mercer, and Sylvia Rojas-Drummond, "Language for the Social Construction of Knowledge: Comparing Classroom Talk in Mexican Preschools," *Language and Education* 13, no. 2 (1999): 133–50.
6. Cindy E. Hmelo-Silver, Ravit Golan Duncan, and Clark A. Chinn, "Scaffolding and Achievement in Problem-Based and Inquiry Learning: A Response to Kirschner, Sweller, and Clark (2006)," *Educational Psychologist* 42, no. 2 (2007): 99–107, doi:10.1080/00461520701263368; William A. Sandoval and Brian J. Reiser, "Explanation-Driven Inquiry: Integrating Conceptual and Epistemic Scaffolds for Scientific Inquiry," *Science Education* 88, no. 3 (2004): 345–72; Aida Walqui and Leo Van Lier, *Scaffolding the Academic Success of Adolescent English Language Learners: A Pedagogy of Promise* (San Francisco: WestEd, 2010).
7. Joshua A. Danish and Noel Enyedy, "Negotiated Representational Mediators: How Young Children Decide What to Include in Their Science Representations," *Science Education* 91, no. 1 (2007): 1–35, doi: 10.1002/sce.20166; Leslie Rupert Herrenkohl and Véronique Mertl, *How Students Come to Be, Know, and Do: A Case for a Broad View of Learning* (Cambridge, UK: Cambridge University Press, 2010); Shirley J. Magnusson and Annemarie Sullivan Palincsar, "Teaching to Promote the Development of Scientific Knowledge and Reasoning About Light at the Elementary School Level," in *How Students Learn: History, Mathematics, and Science in the Classroom*, ed. Suzanne Donovan and John Bransford

(Washington DC: National Academies Press, 2005), 421–74; Josh Radinsky, Sonia Oliva, and Kimberly Alamar, "Camila, the Earth, and the Sun: Constructing an Idea as Shared Intellectual Property," *Journal of Research in Science Teaching* 47, no. 6 (2010): 619–42.

8. Ann L. Brown and Joseph C. Campione, *Psychological Theory and the Design of Innovative Learning Environments: On Procedures, Principles, and Systems* (Mahwah, NJ: Lawrence Erlbaum Associates, 1996); Jim Minstrell and Pam Krause, "Guided Inquiry in Science Classrooms," in *How Students Learn Science in the Classroom*, ed. Suzanne Donovan and John Bransford (Washington, DC: National Academies Press, 2005), 475–514; Passmore and Stewart, "A Modeling Approach"; Marlene Scardamalia and Carl Bereiter, "Knowledge Building: Theory, Pedagogy, Technology," in *The Cambridge Handbook of the Learning Sciences*, ed. Kevin Sawyer (New York: Cambridge University Press, 2006), 97–118; Carol L. Smith et al., "Sixth-Grade Students' Epistemologies of Science: The Impact of School Science Experiences on Epistemological Development," *Cognition and Instruction* 18, no. 3 (2000): 349–422, doi:10.1207/S1532690XCI1803_3.
9. Rick A. Duschl, Heidi A. Schweingruber, and Andrew W. Shouse, eds., *Taking Science to School: Learning and Teaching Science in Grades K-8* (Washington, DC: National Academies Press, 2007); National Research Council, *A Framework for K-12 Science Education: Practices, Crosscutting Concepts, and Core Ideas* (Washington, DC: National Academies Press, 2012).
10. Stéphane Baldi et al., Highlights from PISA 2006: Performance of US 15-Year-Old Students in Science and Mathematics Literacy in an International Context. NCES 2008-016 (Washington, DC: National Center for Education Statistics, 2007); Eric R. Banilower et al., Lessons from a Decade of Mathematics and Science Reform: A Capstone Report for the Local Systemic Change Through Teacher Enhancement Initiative (Chapel Hill, NC: Horizon Research, Inc., 2006); Kathleen Roth and Helen Garnier, "What Science Teaching Looks Like: An International Perspective," Educational Leadership 64, no. 4 (2006): 16; Iris R. Weiss et al., Looking Inside the Classroom (Chapel Hill, NC: Horizon Research Inc., 2003).
11. Rodger Bybee, Barry McCrae, and Robert Laurie, "PISA 2006: An Assessment of Scientific Literacy," *Journal of Research in Science Teaching* 46, no. 8 (2009): 865–83.
12. Richard J. Murnane and Jennifer L. Steele, "What Is the Problem? The Challenge of Providing Effective Teachers for All Children," *The Future of Children* 17, no. 1 (2007): 15–43; Jonah E. Rockoff, "The Impact of Individual Teachers on Student Achievement: Evidence from Panel Data," *American Economic Review* 94, no. 2 (2004): 247–52; William L. Sanders and June C. Rivers, *Cumulative and Residual Effects of Teachers on Future Student Academic Achievement* (Knoxville: University of Tennessee Value-Added Research and Assessment Center, 1996).

Chapter 2

1. Jo Ellen Roseman, Marcia C. Linn, and Mary Koppal, "Characterizing Curriculum Coherence," in *Designing Coherent Science Education: Implications for Curriculum, Instruction, and Policy*, ed. Yael Kali et al. (New York: Teachers College Press, 2008), 13–38; Gilbert A. Valverde and William H. Schmidt, "Greater Expectations: Learning from Other Nations in

the Quest for 'World-Class Standards' in US School Mathematics and Science," *Journal of Curriculum Studies* 32, no. 5 (2000): 651–87.

2. Edward G. Lyon et al., Secondary Science Teaching for English Learners: Developing Supportive and Responsive Learning Contexts for Sense-Making and Language Development (Lanham, MD: Rowman & Littlefield, 2016).

Chapter 3

1. Sarah Michaels and Cathy O'Connor, *Talk Science Primer from TERC's Talk Science Program* (Cambridge, MA: TERC, 2011); Jen Cartier et al., *Five Practices for Orchestrating Task-Based Discussions in Science* (Thousand Oaks, CA: Corwin Press, 2013).

Chapter 4

1. Suzanne Chapin, Cathy O'Connor, and Nancy C. Anderson, *The Inquiry Project: Bridging Research & Practice*, supported by the National Science Foundation (Cambridge, MA: TERC, 2012).
2. Cathy O'Connor, e-mail message to author, October 2016.

Chapter 7

1. Jen Cartier et al., Five Practices for Orchestrating Task-Based Discussions in Science (Thousand Oaks, CA: Corwin Press, 2013).

Chapter 8

1. Researchers also call this *just-in-time instruction*, because the choices of which ideas to expand upon are often driven by students. They feel a need to know more in order to make progress, rather than having the teacher delivering information that they don't know what to do with.
2. Helen Quinn, Okhee Lee, and Guadalupe Valdés, "Language Demands and Opportunities in Relation to Next Generation Science Standards for English Language Learners: What Teachers Need to Know," *Commissioned Papers on Language and Literacy Issues in the Common Core State Standards and Next Generation Science Standards* 94 (2012): 32.
3. Many of the basic principles we've mentioned for this practice can also apply to presentations in which you are teaching a skill (such as graphing, using a microscope, or interpreting a topographical map).

Chapter 9

1. Moving among the tables for this purpose is not a common practice in American classrooms, but it is a staple of teaching in places like Japan. Japanese teachers even have a name for it—*kikanjunshi*. This naming is noteworthy because it signals to professionals that there are strategies for doing this specialized work, and that they can talk together about what they do during this practice and how to make their interactions with students more productive.

2. William J. Ripple and Robert L. Beschta, "Trophic Cascades in Yellowstone: The First 15 Years After Wolf Reintroduction (supplementary data)," *Biological Conservation* 145, no. 1 (2012): 205–13, http://ir.library.oregonstate.edu/xmlui/handle/1957/20842.

Chapter 11

1. We acknowledge that other mechanisms are at play in these controversies, from political interests to antiscience sentiments to suspicion of government or academic authority in general.
2. Some science education researchers use a broader definition for *claim* than we have described here. For them, claims include position statements about social values in science. For example, one might make the claim that animals should not be kept in zoos or that countries should invest more in renewable forms of energy. These types of statements and the questions they generate are very much worth studying. Arguments for or against can be shaped by information and logical reasoning, but they are also influenced by values. These claims are about "what should be done" rather than explaining why something happens.

Chapter 12

1. During the modeling process, too, students can often talk at length about their theories. Yet the simple requests from teachers—"Okay, that's great; can you put that in writing?"—is more difficult than imagined. The written explanation that goes with the models often requires extra scaffolding in the form of graphic organizers or sentence starters: "We think that [this part of our model] may influence [this part]. We think this because _____." Peer-to-peer conversations help generate ideas that should be included in the written explanations and identify academic language that might be included.
2. We are aware that there are differences between theories and hypotheses in science, but we have used "mini-theories" and hypotheses interchangeably in our classrooms. Many young learners are visibly energized when we highlight their ideas with either of these labels. In one of the kindergarten classes we've mentioned previously, where students were exploring "How can someone little bump someone big off the end of a playground slide?" a young girl came up with the hypothesis that "Gravity is everywhere, even at the top and bottom of the slide." While this assertion does not fit neatly into any category of scientific knowledge, we added this to our list of mini-theories early in the unit, and she came to realize that she could generate science ideas—that *her ideas* were *science ideas*. This belief becomes more common over time in AST classrooms.
3. Item #2 (is it filled with steam?) is, in fact, a question that remains unresolved by all our teachers and students who have tried to create complete explanations for this event. No one really knows if the steam is *added* to the gases already in the tanker, or if the steam *drives out* other gases and *replaces* these before the hatch is closed. These two scenarios lead to different causal stories for the implosion.

Chapter 13

1. Anthony S. Bryk et al., Learning to Improve: How America's Schools Can Get Better at Getting Better (Cambridge, MA: Harvard Education Press, 2015).
2. Paul Cobb and Kara Jackson, "Supporting Teachers' Use of Research-Based Instructional Sequences," *ZDM* 47, no. 6 (2015): 1027–38; Elizabeth A. van Es et al., "A Framework for the Facilitation of Teachers' Analysis of Video," *Journal of Teacher Education* 65, no. 4 (2014): 340–56.
3. Joan Richardson, "Norms Put the 'Golden Rule' into Practice for Groups," *Tools for Schools*, August/September 1999.
4. David Yeager et al., *Practical Measurement* (Palo Alto, CA: Carnegie Foundation for the Advancement of Teaching, 2013), https://www.carnegiefoundation.org/wp-content/uploads/2013/12/Practical_Measurement.pdf.
5. Andrew Morozov et al., "Emotional Engagement in Agentive Science Learning Environments" (paper presented at meeting of the International Conference of the Learning Sciences, Boulder, Colorado, June 2014).
6. *Acknowledgments.* The partnership with local schools and districts described in this chapter was supported by grants from the National Science Foundation (DRL 1315995, Building Capacity for the Next Generation Science Standards through Networked Improvement Communities), from Washington STEM (Developing Networked Improvement Communities for High Quality Mathematics and Science Teaching), and from a Washington State MSP Grant (PASTL: Partnership for Ambitious Science Teacher Leaders). Several lead researchers supported the research and development work, including: Dr. Jen Richards, Dr. Karin Lohwasser, Dr. Sara Hagenah, Dr. Christine Chew, Dr. Kat Laxton, Carolyn Colley, Soo-Yean Shim, Christie Barchenger, Bethany Sjoberg, Ann Morris, Ramona Grove, and Dana Dyer. This team of researchers and practitioners created tools and processes that supported PLCs and the diffusion of knowledge across schools.

Appendix A

1. Michelle E. Alvarez and Corrine Anderson-Ketchmark, "Danielson's Framework for Teaching," *Children & Schools* 33, no. 1 (2011): 61; National Board for Professional Teaching Standards, *Science Standards, Third Edition* (Arlington, VA: NBPTS, 2014), http://www.nbpts.org/wp-content/uploads/EAYA-SCIENCE.pdf; edTPA, *2013 edTPA Field Test: Summary Report* (Stanford, CA: Stanford Center for Assessment, Learning, and Equity, 2013).

About the Authors

MARK WINDSCHITL is a professor of Teaching, Learning & Curriculum at the University of Washington. He taught secondary science for thirteen years in the Midwest before receiving his doctorate and moving to Seattle. His research focuses on how teachers take up new practices and the tools they use to engage students in authentic disciplinary activity. Dr. Windschitl is the lead author of "Rigor and Equity by Design: Seeking a Core of Practices for the Science Education Community," a chapter in the newest edition of the *Handbook of Research on Teaching* (American Education Research Association). He is a past recipient of the AERA Presidential Award for Best Review of Research, and a member of the National Research Council Committee on Strengthening and Sustaining Teachers.

JESSICA THOMPSON is an associate professor in Teaching, Learning & Curriculum at the University of Washington. Her research focuses on building Local Improvement Networks that support ambitious and equitable teaching practice with novice and experienced science teachers, science and English learner (EL) coaches, principals, and district leadership. She has expertise in facilitating and studying teacher learning of Ambitious Science Teaching practices at the elementary and secondary level, as well as in the methods of improvement science. Central to her work is partnering with culturally and linguistically diverse student populations in formal and informal settings. She also runs and studies afterschool programs that learn from and support ethnic minority girls' engagement in scientific inquiry. Dr. Thompson has a background in biology and chemistry. She taught grades 6–12 science as well as in a dropout prevention program for eight years in North Carolina and Washington State. At the University of Washington, she teaches secondary and elementary science teaching methods courses, Teacher Learning and School Change, and Culturally Responsive Math and Science Teaching.

MELISSA BRAATEN is an assistant professor of Science Education at the University of Colorado in Boulder. She taught upper elementary, middle, and high school science for thirteen years in Texas and in South Seattle before receiving her doctorate. Her research focuses on the complexities of teaching science in culturally sustaining and responsive ways that disrupt injustices and advocate for justice. In research partnerships with teachers, she drew upon teachers' expertise and insights to refine professional learning experiences across their career trajectory and build stronger explanations of how teachers learn. She is interested in how teaching is shaped by—and how teachers could shape—the political and institutional contexts of schools, educational reforms, and education policy. In 2011, she received the Outstanding Doctoral Research Award from the National Association for Research on Science Teaching.

Index